T0075629

THE WEIGHT OF GOLD

MINING AND SOCIETY SERIES

Eric Nystrom, Arizona State University, *Series Editor*

Our world is a mined world, as the bumper-sticker phrase "If it isn't grown, it has to be mined" reminds us. Attempting to understand the material basis of our modern culture requires an understanding of those materials in their raw state and the human effort needed to wrest them from the earth and transform them into goods. Mining thus stands at the center of important historical and contemporary questions about labor, environment, race, culture, and technology, which makes it a fruitful perspective from which to pursue meaningful inquiry at scales from local to global.

Books published in the series examine the effects of mining on society in the broadest sense. The series covers all forms of mining in all places and times, building from existing press strengths in mining in the American West to encompass comparative, transnational, and international topics. By not limiting its geographic scope to a single region or product, the series helps scholars forge connections between mining practices and individual sites, moving toward broader analyses of the global mining industry in its full historical and global context.

Seeing Underground: Maps, Models, and Mining Engineering in America
by Eric C. Nystrom

Historical Archaeology in the Cortez Mining District: Under the Nevada Giant
by Erich Obermayr and Robert W. McQueen

Mining the Borderlands: Industry, Capital, and the Emergence of Engineers in the Southwest Territories, 1855–1910
by Sarah E. M. Grossman

The City That Ate Itself: Butte, Montana and Its Expanding Berkeley Pit
by Brian James Leech

One Shot for Gold: Developing a Modern Mine in Northern California
by Eleanor Herz Swent

The Weight of Gold: Mining and the Environment in Ontario, Canada, 1909–1929
by Mica Jorgenson

The Weight of Gold

Mining and the Environment in Ontario, Canada, 1909–1929

Mica Jorgenson

University of Nevada Press | *Reno & Las Vegas*

University of Nevada Press | Reno, Nevada 89557 USA
www.unpress.nevada.edu

Manufactured in the United States of America

FIRST PRINTING

Cover design by Louise OFarrell
Cover photograph: "Porcupine City," CPC-03158, Canadian Postcard Collection, The William Ready Division of Archives and Research Collections, McMaster University Library.

ISBN 978-1-64779-104-9 (cloth) • ISBN 978-1-64779-105-6 (ebook)
Library of Congress Cataloging-in-Publication data is on file.

The paper used in this book meets the requirements of American National Standard for Information Sciences—Permanence of Paper for Printed Library Materials, ANSI/NISO Z39.48-1992 (R2002).

Contents

Preface

Mining Stories

Late in August 2017 I stood on an old gold mine's abandoned waste heap in the middle of a classic Canadian Shield landscape. I was at the heart of the Abitibi gold belt near Timmins, Ontario. Stunted alders, willow, and a few unhappy spruce trees lapped at the edges of the stacked tailings, fluorescent green against the rust-colored soil. I could see the enormous berm of Goldcorp's active open-pit mine in the distance. Closer to me, oxidized red water pooled over an old quad road. Abandoned machinery poked up here and there, decomposing quietly into the mud. I had picked some wild blueberries, and now I held them in one hand. I was wondering if they were safe to eat.

I had come to the Abitibi with questions about gold mining. I found the answers piled up in disorganized layers on the land. Goldcorp's Porcupine pit is the longest continuously producing gold mine in North America. It is a gold rush that never ended. From the top of the tailings pile, I could see the whole of its history scattered messily across the muskeg. A hundred years ago, the oxidized sand underfoot had once been solid granite and quartz holding tiny specs of low-grade gold under a patchwork of lakes and forests. These rocks made the Abitibi the heart of a powerful extractive industry. The 1909 Porcupine gold rush drew mining expertise from around the world, and, as Canada's reputation grew on the international stage, it became a node from which mining power, people, and products spread. Now, pieces of that industry's greatest successes and most catastrophic failures are evident on the earth. The land tells the story of people, companies, and states and their search for wealth. Gold mining permeates everything, from the massive geological formations that hold up the continent to the little wild blueberries.

Like a tailings heap, I felt simultaneously at home and out of place on Abitibi's surface. I grew up in the gold mining town of Wells, British Columbia, one thousand miles to the west of Abitibi. In 1862 the Barkerville gold rush turned my quiet subalpine valley into a booming extractive metropolis. By 1900 everyone was gone, but in 1930 the town bustled again as rising gold prices stimulated a second (more industrial) rush. Eighty years after that, my parents took me up overgrown mining roads to access blueberry patches. As we filled our buckets, my dad would point to an old sluice rotting into the loam. "They had hydraulic races that brought the water all the way from the alpine," he explained with wonder, fingers stained blue. There was a kind of pride mixed with horror in the knowledge of what had come before. We knew intellectually that miners used mercury and cyanide to separate gold from bedrock, but we did not worry about whether it had spilled on the soil that fed our blueberries. That was in the past. So as I listened to my dad's stories and ate fistfuls of fruit, I ingested mining history. I took it inside myself where it plunked softly into my cells, like blueberries into a bucket.

As I moved away from Wells for school, the price of gold steadily rose. Sometime in the early 2000s, mining companies speculated the abandoned deposits in Wells might be profitably worked again. Unfamiliar trucks rattled down Wells's only street. As the eyes of outsiders increasingly focused on my hometown, surveys showed dangerous concentrations of contamination from historic mining on the old baseball diamond, on the banks of the river where we swam, and up on the mountain where the berries grew. The provincial government declared parts of my home a contaminated site. Yellow gates went up across old forestry roads, and warning signs popped up like willows from under spring snow. Then a mining company with a series of evolving collection of backers announced the discovery of profitable veins yet to be exploited in the hillsides. This new mining company promised to clean up some of the old waste as part of a brand-new sixteen-year gold mine that would bring jobs to locals and wealth to the province.

When I came home for summer vacations, I found a new infrastructure of toxicity lurking in the bush, haunting the familiar places I had known as a kid. As the land I had known as a child transformed under the blades of bulldozers, I became disoriented. I jumped over mine gates and went into the woods looking for berries, but sometimes I got lost on new roads that cut through once-familiar topographies. As a child I had feared running into black bears, but now I feared getting caught trespassing by strangers in yellow safety vests. I was also plagued by new questions. Was it safe to eat food from mined land? Could a place be both simultaneously life-giving and toxic? Did a mining tenure held by distant people and their international investors invalidate my connection to a place? The transformation occurring before my eyes felt out of my

View from the Buffalo Ankerite Tailings Pile, Porcupine Lake, Ontario. Photo by the author, August 18, 2016.

control, driven by external forces I did not fully understand. I wanted to run time backward. I wanted to peel back the layers of history that had settled on my hometown as heavy and thick as ice-age sediment. I wanted to find the genesis point for this inevitable trajectory that seemed to dictate the fate of all gold-bearing lands, from discovery to contamination, over and over again.

That is how I ended up standing on a tailings pile in the Abitibi, a thousand miles from home, in the summer of 2017. I had begun reading books about mining in graduate school. I had learned that historians have explained it in different ways. Mining towns like Wells (and, as I discovered, like Timmins) have been variously lauded as examples of environmental declension, celebratory nationalism, sacrifice zones, liberal individualism, and the follies of industrial capitalism. As I sought to understand how the past had unfolded into the present, I realized time could not be laid out in an orderly line that started with a nugget in a prospector's palm and ended with a toxic tailings pile. There was something missing. Grand historical theories could not fully explain my experience, or the experiences of others living in mining towns. For one thing, no one mentioned blueberries. From inside my mind and body, gold's legacy had quietly created the scaffolding for a worldview where poison lay side by side with nourishment and where even the most ravaged, broken things gave back in abundance if given time to heal.

The Weight of Gold is my effort to make sense of gold mining's legacy, from the sprawling structures that lace the underground to the cells that make up our bodies. It is my explanation for how mining converges on our communities, why its past continues to shape our extractive present, and how we might begin to care for damaged places.

Acknowledgments

This book is the product of more than thirteen years under the tutelage of many generous scholars and friends. Its genesis lies on Dakeł and Secwépemc land, deep in the Cariboo Mountains of British Columbia, where questions about gold and the environment first began to trouble my mind, and extends to Treaty Nine land belonging to Mattagami First Nation. Thank you to the family, friends, neighbors, and community members living with gold mines old and new for your curiosity and enthusiasm in response to my many questions. I am humbled by the way that my research has been adopted, improved upon, and wielded by those grappling with the modern impacts of historic gold mining.

At the University of Northern British Columbia, Jacqueline Holler, Jon Swainger, and Ted Binnema were the first to encourage my enthusiasm for environmental history. Some combination of sixteenth-century Mexican court records, Herodotus's *Histories,* and microfilmed Hudson's Bay Company (HBC) journals made me choose history and graduate school. Later at McMaster University, Michael Egan, Ken Cruickshank, and John Weaver gave me a stack of books by environmental historians that cemented my obsession. After I finished reading, they patiently worked with me to create the dissertation that was the foundation for this book.

During my PhD process, Liza Piper and Viv Nelles taught me not to be afraid of writing big history. Over coffee, panels, and book tables, countless environmental historians at the American Society for Environmental History provided helpful feedback and energy as I developed and refined my ideas. I am grateful to have found an intellectual home with the ASEH. I am also grateful to the Mining History Association, and especially to Brian Leech and Eric Nystrom who made me feel welcome, supported my work, and were always game to join my panels. Without them, this book would not exist. John Sandlos and Arn Keeling have also been instrumental. They provided valuable mentorship at different times throughout my work and provided considerable intellectual support through workshops, over coffee, and via many email exchanges.

I am grateful that my path through the academy intersected at various times with those of Andrew Marcille, Bre Lester, Blake Bouchard, Chelsea Barranger, Heather Green, Hereward Longley, John Baeton, Lorena Campuzano-Duque, Nevcihan Ozbilge, Samantha Clarke, Scott

Johnston, and Spirit Waite. Among the many wonderful colleagues I have been privileged to work with over the years, they have endured my overenthusiasm for mining history more often than most. They were always there to share my excitement about things no one else cared about, commiserated with me about the struggles of our shared work, and otherwise spent a lot of time talking together when we all could have been writing. Our relationships began in graduate school but have only gotten stronger in the years after. I aspire to be half as smart, talented, and generous as any one of them. I miss our shared offices because I miss being around them every day.

Thank you to Mattagami and Matachewan First Nations, whose lands I visited regularly during my research in Porcupine and Timmins. Thanks also to the Archives of Ontario, Library and Archives Canada, the Timmins Public Library (especially Karen Bachmannand), the Queen's University Archives, the J. N. Desmarais Library, the Cobalt Museum, the McMaster Map's library (especially Jay Brodeur), and the William Ready Division of Archives (especially Bridget Whittle). Working with archival documents is the best part of any historical project, and the work of dedicated archivists makes all historical research possible. My work was funded at various stages by the generous support of the Social Sciences and Humanities Research Council of Canada, Wilson Institute for Canadian History, James Robertson Carruthers Memorial Award, Ross Reeves Grant, Armstrong History Fund, and Mining History Association. The bulk of this manuscript was written during my time at the Lewis and Ruth Centre for Digital Scholarship, where I worked under the guidance of Dr. Andrea Zeffiro and Dr. Dale Askey.

My chosen family, Eric Lane, Alexandra Lane, Cole Mueller, Katie Cornish, Erin Kleven, Jill Hughes, and Sarah Lawley-Wakelin, have been with me from the beginning, from British Columbia to Gangwondo to Ontario to Rogaland. Erin even read my entire dissertation. Thank you to my mom, Shirley-Ann Royer; my brother, Ekai Jorgenson; my grandmother Pauline Royer; and my in-laws, John and Cathy Lane, for supporting me unconditionally over all these years.

It has been a pleasure to work with the publication staff at the University of Nevada Press, particularly Margaret Dalrymple, JoAnne Banducci, Annette Wenda, Curtis Vickers, and Jinni Fontana. Special thanks also to the four anonymous readers who provided feedback that has greatly improved the book. Any remaining errors are despite their best efforts.

THE WEIGHT OF GOLD

Introduction

The Mining Environment at Porcupine Lake

If you ever have a chance to ask the old-timers about the Porcupine gold rush, they will tell you that its discovery was destiny. In an article titled "Mountain of Gold Found by Fluke," Chicago businessman and early investor William Edwards listed the series of fateful events that led to the discovery of his company's "golden staircase" on the shores of Porcupine Lake. Prospectors had dug within meters of the deposit and found nothing. An "Indian" had told another party that there was gold nearby but had been ignored. Eventually, a prospecting party had pounded some stakes in the ground at random and in doing so unwittingly marked out the ground that would become the great Dome Mining Company. In this vision of progress, human exploitation of gold was fated by God to ensure the proper utilization of the land: "A curious thing about [gold] is, that it has been deposited largely in wild, inaccessible, desolate regions. One could imagine that [gold] had been purposefully hidden away to call men to go forth and traverse the earth."[1]

Do not believe this version of the past. The first finds were instigated by a collection of men who went into the Abitibi with the specific intention of staking mineral claims at the behest of a government determined to make the North a productive hinterland. The Porcupine gold rush happened on a busy portage route carrying local people between the Mattagami River and Frederick House Lake, an important location of hunting and farming near an old fur-trade fort. To the south, the waning opportunities offered by a fading silver rush had sent prospectors fanning out along well-established Anishinaabe transportation routes described by government surveyors in 1909. In the cities, a whole class of men with capital had been primed for investment by similar discoveries, most recently in Alaska and the Yukon. So when moss scraped from a quartz outcropping near Porcupine Lake revealed high-grade gold, the world knew what to do next. After all, no one could resist a deposit that, in the words of its discoverers, looked "as though someone had dripped a candle along it, but instead of wax it was gold."[2] Prospectors pulled the topsoil away from their dripping vein then paddled the 132 miles back to the nearest telegraph office to tell their American financiers the good news. As the snow began to fall and Porcupine Lake began to freeze, the area became a maze of staked claims and disturbed dirt. Companies formed, money was raised, and a steady stream of people and equipment poured onto the land. The Porcupine gold rush was on—not because of destiny or luck but because of history.

For an industry that would later credit its success to fortune, the land around Porcupine Lake took an incredible amount of work to be made productive. Every ounce of gold that came out of the bedrock over the next century (more than sixty-seven million ounces) had to be wrung from an industrial environment pushed regularly over the edge of viability. The lake that gave the first rush its name is small and unassuming, one of thousands of similar water-filled depressions dotting the Canadian Shield. Yet the bedrock stretching underneath possessed geological conditions capable of producing and sustaining mining. There was enough surface gold to create a first rush of excitement and enough low-grade gold to keep those with capital and expertise profiting over time. The quick victories of those initial finds created foundations of capital, infrastructure, expertise, and labor that in turn enabled larger-scale and more expensive methods of extraction to come later. In the years after 1909, people dug deep into the rock, altered bodies of water, and cut down every single tree for miles in every direction. They erected tents and then built dusty rows of houses for their families. They cut in roads, bridges, and hydroelectric dams. Then in 1911 they built everything all over again after forest fires destroyed it. Miles below the surface, people and machinery crawled through a honeycomb of tunnels and caverns. Every single pebble was turned over, fed through mills, and put back down in new configurations. The power required for such massive operations meant flooding the river valleys above the mines where people had lived. As the underground tunnels and lakebeds filled with waste, they became explosive and toxic hazards. This was the cost of keeping gold coming out of the ground.

Both the environmental challenges and their solutions at Porcupine stemmed from the complex ecology of humans, machines, and land built over time.[3] Although production occasionally waned, new industrial techniques (often generated on other goldfields) could be applied to the land to ensure continuing profitability. Those techniques functioned by transferring the costs of extraction to places where people could not see them. To this day, when exploration companies in northern Ontario dig through layers of old shafts and turn ever-smaller concentrations of Abitibi gold into dollars, they are extending an infrastructure rooted in the early twentieth century. They are finding new ways to defer mining's costs. Porcupine's history is one of a series of crises mitigated through a combination of relentless optimism on behalf of a few key players, the hard physical labor of local people, and the timely interventions of a growing international extraction economy increasingly connected to the land through both material and immaterial networks.

Porcupine is not unique. Neither its geology nor its society was notably distinct from similar goldfields the world over. In fact, anyone who comes from a mining town anywhere in the world has likely heard a similar story of environmental change to the one I have just described. This is because the world's gold mines are closely related to each other. Porcupine's power is partly in the way that its history conforms to many of the same patterns that shaped mined landscapes in Australia, New Zealand, South Africa, the United States, Latin America, and Europe. It is part of a series of similar discoveries following predictable social and ecological patterns on multiple continents. Despite unique cultural and ecological conditions, mining communities the world over carry pieces of a shared extractive history.[4] In general, gold rushes like the one that sparked mining at Porcupine tend to occur far from urban centers during times of economic growth, when individuals and corporations are more risk tolerant and open to new opportunities. The California rush in 1848, Australian rush in 1851, British Columbia rush in 1858, and Otago rush in 1861 all came at times of economic expansion (the Klondike gold rush of 1896 is an exception, generated at a time of economic depression and industrial malaise when gold promised relative stability, freedom, and profit).[5] Prospectors are usually migrant male members of a neo-European culture sharing specific ideas about their relationship to nature.[6] Discoveries trigger media excitement, speculation booms, migration, expropriation of land and resources, and infrastructure development. Running parallel to these changes are a series of environmental crises and mitigation efforts: lack of governance, fires, floods, explosions, or disease. From the beginning, most mines are haunted by the specter of resource exhaustion, which eventually ends all extractive ventures. This basic chronology played out fully at Porcupine: a brief prospecting period based on surface gold quickly gave way to syndicate-funded development, expansion underground, community building, large-scale environmental change, and, eventually, remediation efforts. What makes Porcupine useful for study among its mining relatives is the extended nature of its trajectory. Where other gold mines lived and died (and sometimes lived again), Porcupine *endured.* Among the "big three" mining companies that came to dominate the district, McIntyre Mine ceased production in the late 1980s, Dome closed in 2017, and the Hollinger pit is, as of writing, still operating. Pieces of other mining stories are visible at Porcupine, but they are sometimes broken or interrupted in ways that Porcupine's story is not.

It is especially important to understand Porcupine's place in the context of international mining because Porcupine occurred at the crux of a wider industrial transition in mining in the early twentieth century. In 1909 Porcupine became part of an exchange that continues to occur between mining states whereby extractive projects benefited from economic, political, and cultural networks.[7] The turn of the century was a moment distinct from the high-grade placer gold mining exemplified by the early days of the California, Australia, British Columbia, and Klondike gold rushes. As surface gold became harder to find, many of these mining fields experimented with capital-intensive dredging and hydraulic technologies where local conditions made them viable. Porcupine occurred at a time when industrial techniques that worked best on microscopic gold and required specialized knowledge, equipment, and capital to extract were becoming increasingly widespread.[8] Internationally, the industry had begun to organize around new types of deposits in response to a variety of factors, including the rising price of metals, the availability of new technology, declining placer deposits, and a supportive political and economic climate. At precisely this moment in the international mining milieus, the provincial government in Ontario decided to turn its northern reaches into a profitable hinterland. Porcupine was just the start. Eventually, northern Ontario, Quebec, and Manitoba would be peppered with burgeoning metal mines the length of the Abitibi gold belt that continue to dig up profits to the present. By the end of the 1920s, Porcupine's successors at Kirkland Lake, Red Lake, and, later, Rouyn-Noranda and Val-d'Or would outstrip it in terms of size and production values.

Porcupine's story shows how history flowed easily between mining states. As a node for international mining networks between 1909 and 1929, Porcupine became part of the foundation for industrial regimes that continue to shape mining communities in the present. Within this wider context, Porcupine's resemblance to other mining fields is no coincidence: when news of gold at Porcupine broke, the Ontario government looked to its recent predecessors in British Columbia, California, and Witwatersrand, as it pivoted expertly to support a northern mining economy. Although working within a unique Ontario legal tradition, politicians wrote Ontario mining law with explicit reference to the experiences of their international counterparts. Hoping to attract and encourage prospectors in 1864, for example, the Ontario cabinet did away with royalties and set the outright price of ownership of mineral-bearing land at two dollars per acre,[9] citing similar measures during "the Californian discoveries in 1848, those of Australia in 1851, and of British Columbia in 1858."[10] This legal structure would go on to encourage a mining regime friendly to the international syndicates who paid the first prospectors—many of whom had already tried their luck on other goldfields before arriving at Porcupine. Porcupine's major lodes were discovered by people who had mined in the United States and California before arriving in northern Canada and used technology invented, developed, and manufactured in older goldfields to extract gold from Ontario bedrock. Much of Canada's modern methods for gold mining would be imported and tested at Porcupine.

Porcupine's history was not a passive one-way flow: incoming forces interacted with northern Ontario's environment in unpredictable ways. Geologists and engineers exchanged blueprints, scientific research, and expert knowledge across continents. They experimented on the shores of Porcupine Lake and then shared what they had learned with their contemporaries on other mining fields. Canadians moved frequently across borders, especially the forty-ninth parallel, and occupied important positions in the industry. For example, Edward P. Mathewson was educated at McGill before working in a variety of positions that took him from Colorado to Montana, Mexico, Ontario, New York, and Arizona. When he became president of the American Institute of Mining and Metallurgy in 1923, he was mistakenly called the first Canadian to hold the position until it was pointed out that, in fact, he had been preceded by a collection of Canadian mining professionals with close ties to the larger international milieus. The *Engineering and Mining Journal* acknowledged the influence of Canadian mining professionals in the field: "Our excellent neighbors of the North have given the American mining and metallurgical profession a number of men of the first rank."[11] Mining history simultaneously converged at and spread outward from Porcupine, which became a temporary metropole for extractive experience at a moment of transition in the industry as a whole.

As much as the events that followed the discovery of gold on the shores of Porcupine Lake were well rehearsed, progress did not go smoothly. If the turn of the century was a moment of transition for gold mining, it was also a moment of uncertainty and some angst. In 1909 many of the world's mining economies suffered from deep disillusionment about the costs and benefits of the craft. If a few lucky men had profited, the gold rushes of the nineteenth century had largely failed to deliver on the promises of wealth under liberal individualism. Even in places where accidents of geography produced large, dependable hard-rock deposits, any sense of long-term stability remained stubbornly unattainable. Witwatersrand's predictable "reef" formation allowed for rapid takeover by industrial developers, but environmental conditions and tensions with farmers, pastoralists, and migrant labor plagued the industry.[12] In the Klondike, exuberance over new discoveries was tarnished by food shortages and legal difficulties exacerbated by the goldfield's severe winters and isolation.[13] The decline of prominent goldfields in California, Australia, and British Columbia combined with several wildcat speculation busts created pessimism among investors despite the excitement provoked by new finds. In part, industrial mining promised an end to the unpredictability that had plagued the volatile rushes of the nineteenth century. In 1909 Canada had yet to see evidence of its efficacy.[14]

By the end of the 1920s, the mines at Porcupine reached their apogee. They began to look like proof that gold mining could produce healthy and sustainable economies. Mining companies and their advocates framed industrial extraction as the natural end point of the nineteenth-century gold rushes. They drew a straight line from the image of the lone prospector in the Ontario bush to their productive industrial mines and celebrated their accomplishments in the face of the odds. In 1928, conscious of Porcupine's place in a narrative of progress, Ontario minister of mines Charles McRae called Canada "the mineral treasure house to

[the] world."[15] If you followed the gold that came out of Porcupine rock, it would take you all over the globe. It went to pay out investors in Toronto, New York, Johannesburg, and a dozen other places. It funded Canada's allies in the First World War. It enabled major advances in medicine and geology and engineering. It bought diamond drills and railcars and miners' labor. Canada's industrial ascendance was forged over two decades of interaction between humans and nature in northern Ontario's shield landscape. It came out of several catastrophic failures, partial technological solutions, and some lucrative success stories. These experiences granted McRae and his mining contemporaries the power and the confidence to extend the lessons they had learned at Porcupine into the world through the networks they had built over time.

Porcupine's legacy is one of notable stability in an industry famous for boom and bust, but its longevity came at a significant cost. If one side of Porcupine's story is extractive triumph, the other is of a mining camp regularly confronted with existential crisis. Even after Indigenous title had been (theoretically) cleared and the most important claims staked, geologists and prospectors struggled to make Abitibi's twisted, unpredictable rock work for them. A wildfire and a world war destroyed infrastructure, cut supply lines, and reduced available labor. Only the biggest mining companies survived these trials, but as those companies expanded, they encountered new problems associated with their scale. Industrial working conditions in the mines exposed some groups of people to greater risk than others. The giant mines also took up more and more space on the land, through waste disposal and new hydroelectricity projects. They created a landscape that, today, is pocked with pit mines, honeycombed with tunnels, and covered in waste. In the twenty-first century, people in the Abitibi and all over the world are still learning how to live with Porcupine's legacy on the land and in their bodies.

In the following chapters, I make two arguments. First, northern Ontario participated in an international exchange of extractive knowledge, technology, and culture between 1909 and 1929. Porcupine possessed (and in some cases generated) many of the patterns that persist in the behavior and rhetoric employed by the mining industry to overcome the limits imposed by its environment. The chapters describe a series of environmental crises for mining: its remoteness and difficult character, a forest fire, water shortages, waste mismanagement, and industrial disease. Each crisis exposed connectivity across international borders. My second argument is that these environmental crises and their solutions show how Canadian mining's continued profitability depended on the ability of mining companies to redistribute the burdens of mining onto the land and onto marginalized people—a foundation on which the industry continues to rest both at home and abroad.

The Weight of Gold is part of the long tradition of telling mining stories—by old-timers and almost everyone else. The mining-history genre classically served internally focused celebratory nationalist mythologies that rarely referenced other mining states. According to this mythology, gold rushes and gold mining provided the unique and essential cultural, political, and economic conditions for independent state building in Australia, Canada, New Zealand, South Africa, and the United States.[16] For example, mining and prospector myths are part of the "frontier thesis" popularized by Frederick Jackson Turner in the United States.[17] In Australia, Geoffrey Blainey argued that the gold rush created the historical conditions underpinning the shape of his nation.[18] In Canada, mining played a central role in resource-focused economic and political theories of the genesis of the state by historians like Michael Cross, D. M. LeBourdais, and Harold Innis.[19] It later became central to revisions of the "staples thesis" proposed by Viv Nelles and his contemporaries, whose work continues to provide explanatory frameworks for the shape of the Canadian state. Porcupine specifically has inspired a rich collection of largely celebratory popular and academic work, including Michael Barnes's *Fortunes in the Ground* and Phillip Smith's *Harvest from the Rock*.[20]

Since the environmental movement and the emergence of environmental history in the 1980s, some historians have turned the old celebratory mining myths upside down. These historians argue that if the mining industry granted the state a hardy and enterprising spirit, it also left a legacy of exploitation, destruction, and shortsighted profit seeking. In the hands of this new generation, mining stories became criticisms of the nation and its dysfunctional relationship to nature.[21] In 2010 urban environmental historian Steve Lerner used "sacrifice zones" to think about locations where low-income and racialized communities experienced disproportionate environmental harms related to contamination, environmental degradation, and industrial development. Mining historians have extended this idea to practices like mountaintop removal, contamination, and pit mining that disproportionately affect rural, Indigenous, racialized, and economically marginalized people.[22]

Between celebratory nationalism and environmental declension, the current generation of mining historians has sought to understand mining as a complicated interaction between people and the land. In places like Asbestos, Quebec, and Zacatecas, Mexico, environmental historians point to tension between working-class identities and the environmental injustices associated with life on mined land.[23] The ways people adapted to and worked with the mines give the idea of a "sacrifice zone" a human face. By digging into the lived experiences of people on mined land, these historians have shown how industrial extraction does not preclude attachment, use, and life. At the same time, these stories do not flinch away from the serious harms mining brings to environments and communities.[24] Mining history becomes relational: an extension of Thomas Andrews's "failed relationship" metaphor of mining (adapted from Richard White's *Organic Machine*) or a blurring together of organic and inorganic labor as described by Liza Piper.[25] Some of these authors see mining as part of broader scientific, political, and social systems that extend across borders.[26] In the Anthropocene, mined landscapes offer lessons for building relationships with and across the land during times of disorienting environmental change.

The Weight of Gold extends this recent work. Although historians have gone some distance in situating gold mining firmly in internal networks of industrial capitalism, Porcupine shows how nationally based history underemphasizes the degree to which mining states (and *especially* Canadians) looked across borders for answers.[27] It also shows how local histories based in specific contexts relate intrinsically to wider international trajectories. Scientific and legal networks built between Porcupine and the rest of the world were rooted in material exchanges. Like flecks of low-grade gold in the Abitibi bedrock, rich minutia permeate the stacks of corporate correspondence, government legislation, miners' letters, geological reports, maps, and newspapers. On their own they are of limited value; together they provide profitable indications of larger historical structures. Luckily, accessing them does not require drilling.

But piecing together the minutia is only half the historian's work. The job is complete only once wider forces of capital, knowledge, and political power are related back to the gold dust in the rock and the individual people that give them meaning. When the mining industry encountered challenges imposed by the nature of the land it worked, conflict with other resource users, and environmental crisis, they solved them by transferring their costs. At the local level, heavy-metal pollution, appropriation of land, and pit mining disproportionately affected Porcupine's working-class and Indigenous residents. The industry also began deferring problems to the future or to marginalized people in South America or South Africa, where they could not easily be quantified. Mining was a slow violence in which toxicity and damage accumulated gradually and then worked its way outward from extraction sites.[28] Tying international stories to local ones shows how networks and relationships across time and space are concrete forces with the power to materially shape the world.

In order to answer to its different scales, *The Weight of Gold* combines a place-based microhistory with the tools and breadth of transnational or "big history." It does so by starting from the assumption that the mineral booms of the late nineteenth and early twentieth centuries were part of a continuum of frontierism and resource development across multiple British colonies and successor states. As a product of a long history of extraction, mining at Porcupine reflected, reinforced, and advanced elements of mining history stretching through time and space. In practical terms, the narrative "zooms in" on illustrative local events before "zooming out" to place them in their wider context. As they mixed international practice with local needs at Porcupine, mining companies, governments, and individual miners contributed to the foundation for modern relationships with nonrenewable resources. The history of Porcupine is a history of negotiation between global industrial forces and nonhuman nature writ large. By tracing the way that Canadians imported, exported, and adapted methods of extraction at Porcupine, *The Weight of Gold* shows how industrial landscapes were the result of a wide series of human choices across multiple contexts and environments.

The Weight of Gold will follow all the classic beats of a mining story in roughly chronological order, from the exploration phase to industrial maturity. Each chapter centers a different environmental crisis and the "resolution" that allowed mining to continue while delaying or offsetting its true costs. Chapter 1 is about the colonial government's tenuous understanding of and control over the land, which it solidified through the treaty process and a series of scientific surveys. It draws together the multiple threads of history that produced the gold rush and its myths. It starts with the relationships and people who lived around the mine sites before gold was extracted there and chronicles the changes that came when an international industrial mining economy converged on what had long been a landscape dominated by hunting and small-scale farming. I will forgo the long-outdated mythology of the lone (white, male) prospector who shouts "Eureka!" and founds the colonial settler state. Instead, chapter 1 traces the calculated efforts of the settler state to exert economic control over its hinterland: a project that would never be satisfactorily completed. By the end of 1910, there was no longer any doubt about the depth, richness, and long-term viability of Porcupine's gold deposits. Yet the still-small mining companies emerging out of the rush struggled against each other and against northern Ontario's erratic geology as they attempted to establish themselves at Porcupine. Veins changed direction or ended abruptly, and the low-grade nature of the rock required companies to import technology from the United States and the South African Rand. Although some disenchanted prospectors declared Porcupine a "rich man's game," extraction remained decentralized, relatively unorganized, and uneven. Chapter 1 shows how Porcupine connected to international mining from its genesis and how those connections put the region on a path toward industrial extraction.[29] It focuses on Indigenous land use, Indigenous labor, government surveys, the numbered treaty process, multiple (failed and successful) attempts to mine nearby, the arrival of international capital, the development of new technology in South Africa and California, tension between the United States and Canada, and the relentless international promotion of Ontario's mineral wealth.

Then in the summer of 1911, deadly wildfire ripped through northern Ontario, wiping the nascent mining infrastructure off the face of the rock and killing more than a hundred people. The tenuous railway connection between Porcupine and the world was abruptly severed, leaving survivors isolated. Chapter 2 explores how the Great Fire of 1911 literally and metaphorically cleared the way for the dominance of a large-scale industrial vision for mining based on international examples while exposing the threads of its more informal connections to the world. It rearranged Porcupine's physical character in ways that benefited some stakeholders more than others. Large companies financed by international capital consolidated the properties of their burned-out neighbors to create megamines. They took advantage of the burned-off vegetation to get a better look at the exposed patterns in the shield rock. Environmental problems during and after the Great Fire of 1911 are tangled up with the acceleration of large-scale mining, growing inequality, and increased integration between Porcupine and international neighbors.

Chapter 3 argues that the outbreak of the First World War further consolidated big mining at Porcupine. Smaller companies could not endure the economic slowdown, which exacerbated the financial hardships of the Great Fire. Larger companies, on the other hand, could afford to sit on their reserves, buy up the properties of their struggling neighbors, and concentrate on exploration, geological science, and infrastructure development. Aiding large projects was the fact that the war glorified metal production as a form of patriotism and, as it came to a close, created a large body of migrant ethnic labor, which the Canadian government shuttled into northern Ontario. By 1919 the largest mining companies emerged from the war economy more successful, confident, supported, and informed than ever. With the power of large-scale extraction, international capital, and consolidated industrial technology, the Porcupine mines became major competitors on the international stage. Yet as they entered the Roaring Twenties, these companies increasingly chafed against environmental limits. Despite early optimism, shortages of hydroelectric power stalled extraction and cut into bottom lines. Solving the problem meant building bigger reservoirs and, in the process, dispossessing Indigenous people. The war and the water shortages demonstrate how the convergence of international and local forces inscribed physical change on the land and its occupants.

Chapter 4 explores some of the unexpected side effects of industrial expansion. It starts with a bang. Gold mines are not supposed to explode, but in 1928 this thinking changed as flames and toxic gas ripped through Hollinger mine. In the aftermath, Hollinger became the center of a discussion around mine rescue and the impetus for legislative change in the United States. Chapter 4 shows how the Hollinger disaster stemmed from the new problem of industrial mine waste. The Hollinger fire was not the only way the problem manifested: several high-profile court cases also came out of Porcupine's waste problem as mines sued each other for space and resources related to disposal of waste rock, garbage, and water. In solving its waste problems, Porcupine contributed meaningfully to international conversations about environmental and safety issues for the first time. At home, the mines began implementing and celebrating a series of conservation projects aimed at protecting animals and land from the side effects of extraction.

Chapter 5 addresses a slower and more insidious consequence of industrialization—silicosis. Before the 1920s, Canadian extractive companies imported international industrial expertise while considering themselves immune to industrial environmental disease. Silicosis had wreaked havoc on miners' bodies and on corporate bottom lines in Australia, South Africa, and the United States, but mining experts believed Canada had safe, silica-free rock and healthy workers. The discovery of silicosis in Canadian miners in 1924 sent the industry into crisis. When international science offered no immediate cure, work began on a Canadian solution. After the 1920s, Canadian companies exported their new expertise in the form of a medicine called "McIntyre Powder." Alongside their cure, they exported a collection of authoritative scientific papers and regulatory recommendations to their mining neighbors. It would take another fifty years before Canada fully acknowledged that McIntyre Powder itself posed a potential threat. By then the "cure" had been administered to hundreds of thousands of miners across multiple mining states.

The market crash of 1929 ushered in a long period of relative stability for Canadian industrial gold mining. Unlike other commodities, gold maintained and even increased in value during the 1930s, and mines enjoyed high availability of labor and abundant electrical power. These factors, in combination with new scientific networks, placed Canada among the industrial giants of the twentieth century in Australia, South Africa, and the United States. Canadians looked back on the last two decades of development with pride, and Porcupine became narratively linked to the rise of a wealthy, successful industrial state.

The conclusion looks beyond the temporal end of the book to connect Porcupine history to what came after. The conclusion outlines the major events of Porcupine's aftermath in the twentieth century, chronicling the dominance of mining in the North and the stories people told about Canadian gold. It briefly outlines the history of Canadian mining as it proceeded over the next century to the present and points to the ways in which the patterns we have seen at Porcupine replicated elsewhere, both at home and abroad, in the years since. *The Weight of Gold* ends with a reflection on what Porcupine can tell us about Canada's modern place on the stage of international mining, how it was built, and the values at its roots.

CHAPTER ONE

"Promise of Reward to the Prospector"

Making Mines Out of Muskeg in Northern Ontario

With the benefit of hindsight, it would be easy to take Bureau of Mines surveyor William Arthur Parks's offhand comment, "I regard the region south of the trail to Porcupine Lake as giving promise of reward to the prospector," as prophetic.[1] The quote appears in his 1899 report, part of a broader survey of northern Ontario, and refers specifically to the portage between the Mattagami River and Nighthawk Lake where the region's richest mines would rise from the muskeg a decade later. It is almost as if Parks could envision the next century unfolding before him: the millions of ounces of gold under his feet, the development of Ontario's North, the central role of the Abitibi gold belt in ensuring Canada a place among the twentieth-century industrial giants, and the enormous pit mine on the very spot where he stood.

Such a story might fit neatly with the older, celebratory narrative of gold discovery. By 1899 the world had witnessed a series of resource and land rushes that transformed former colonial frontiers from western America to Western Australia to the South African Transvaal. The intervening years between the 1850s and 1899 had seen a contraction in the investment market related to economic depression, but the turn of the century saw a favorable return of the global investment climate and the revitalization of exploration activity. In the context of the general course of the nineteenth century, the discovery of mineral wealth in northern Canada could easily be framed as just the latest domino to fall at the end of a long line of predecessors.

A closer examination of Parks's report and those of his contemporaries reveals that there was nothing inevitable or prophetic about mining at Porcupine. The initiation of a mining regime resulted from overlapping contexts on the Canadian Shield, setting the region on a gradual path toward industrial extraction. The Porcupine gold rush, when it happened, came from the confluence of local and external forces that

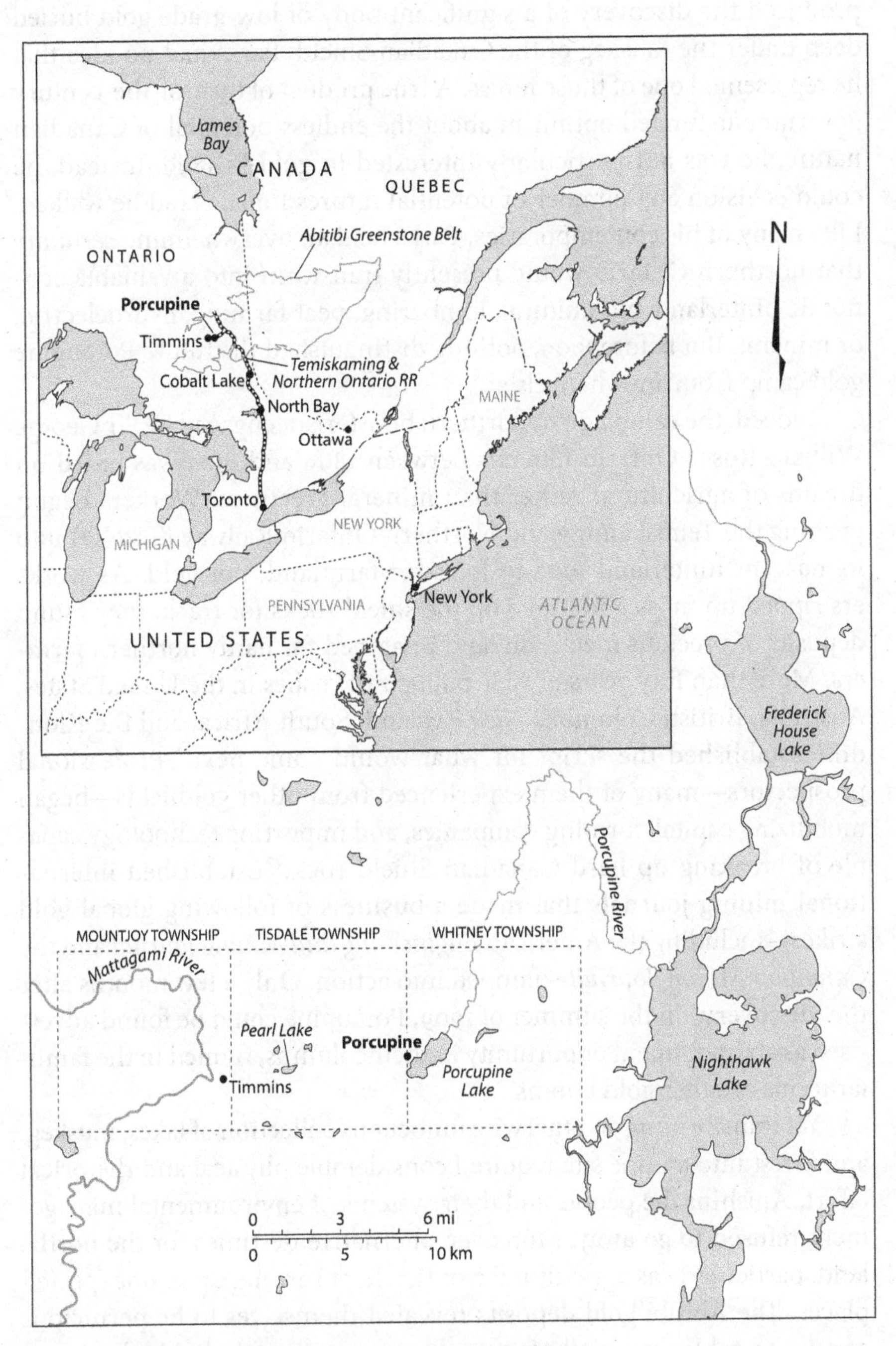

Map of the Porcupine goldfields. Courtesy of Bill Nelson.

produced the discovery of a significant body of low-grade gold buried deep under the muskeg of the Canadian Shield. Parks had no idea that he represented one of those forces. A true product of turn-of-the-century government-funded optimism about the endless potential of Canadian nature, he was not particularly interested in gold in 1899. Instead, he could envision any number of potential futures for the land he walked. Like many of his contemporaries, Parks held an overwhelming *certainty* that northern Ontario would presently transform into a valuable economic hinterland: agricultural, lumbering, peat farming, hydroelectric, or mineral. But before 1909, nothing distinguished the future Porcupine gold camp from anywhere else.[2]

Indeed, the campaign of northern boosterism engaged by Sir George William Ross's Ontario Liberals between 1899 and 1905 was based on dreams of agricultural rather than mineral greatness. Workers begun pushing the Temiskaming and Northern Ontario Railway (T&NO) into its nascent hinterland 1902 to look for farmland, not gold. As workers ripped up moss and broke up the shield rocks for track, they found deposits of precious metals on land imagined for hardy northern farmers. More than fifty years of well-publicized rushes in the United States, Australia, British Columbia, New Zealand, South Africa, and the Klondike established the script for what would come next.[3] Professional prospectors—many of them experienced from other goldfields—began mobilizing capital, forming companies, and importing technology capable of breaking up hard Canadian Shield rock.[4] Established international mining journals that made a business of following global gold strikes—including the American *Engineering and Mining Journal* and the *Canadian Mining Journal*—jumped into action. Only a few months after the "discovery," in the summer of 1909, Porcupine could be found advertised as an investment opportunity in their columns, framed in the familiar terms of other gold booms.

Yet transforming an otherwise innocuous collection of lakes, muskeg, and forest into a mine site required considerable physical and rhetorical effort. Anishinaabe people and their systems of environmental management refused to go away. Moreover, alternative destinies for the northland, particularly as agricultural frontier, kept turning up in unexpected places. The Abitibi gold deposits revealed themselves to be pernicious and inscrutable, especially for small companies or individual miners. Porcupine quickly became a messy convergence of the triple forces of government optimism, local material and cultural geographies, and international mining markets. Insect and fire cycles, vegetative succession, variable precipitation, severe seasonal shifts, and preexisting human hunting, trapping, gathering, agriculture, route making, and other activities shaped the landscape in ways not always conducive to mining.

Additional barriers stemmed from the fact that the early twentieth century was an awkward time for a global mining industry caught halfway between the ideal that any individual could "strike it rich" and the reality of growing corporate dominance. Aside from a few spectacular surface deposits, most of Porcupine's gold existed in complicated low-grade deposits inaccessible to individual miners. Even companies with the ability to recruit international technologies, capital, and labor had trouble applying them. The unique and shifting conditions of the Canadian Shield meant that progress could be uneven, unpredictable, disappointing, and slow. The messy convergence of history at Porcupine Lake facilitated the establishment of the mining industry by 1910, but not without difficulty.

Making a Provincial Hinterland in Anishinaabe Territory Before 1909

The Abitibi's gold deposits remain poorly understood to the present. Even the most up-to-date information on the Ontario Ministry of Northern Development and Mines website compares finding precious metals to looking for "a needle in a haystack" because the deposits are tiny pockets of wealth distributed in unpredictable patterns within a vast and varied body of the Canadian Shield.[5] What we do know is that collisions within Canadian Shield rock between 4.5 and 540 million years ago provided the necessary conditions for gold deposition. These collisions caused cracks or other spaces within host rocks.[6] Gold atoms floating in liquid magma pushed into these spaces and hardened, often within crystalline structures such as quartz and feldspar.[7]

The Canadian Shield contains some of the oldest rocks on earth. Over several million years, cycles of erosion, sedimentation, and tectonic shifts changed the shape and location of Abitibi's gold. Unlike in British Columbia and the Klondike, where the mountains are comparatively young and the gold freshly exposed to the forces of wind and water, much of the gold in Ontario was long ago broken apart and widely dispersed by time and nature. The most recent ice age finally retreated fully from northern Ontario between 8,000 and 9,000 years ago, leaving behind a complex network of lakes and rivers.[8] Some gold deposits remained protected within polished-down outcroppings, but others were swept away and redeposited wherever water and the topography took them.

Above the shield's shifting geology, the land has always been occupied. As the ice retreated north, it left a mixed landscape of low, wet muskeg and higher, glacially eroded shield rock. In general, the soil is thin and acidic (except where the clay belt covers part of the region) and the summers cool. Berry crops thrive, and the entire shield is subject to regular regenerative forest fires, which people used to initiate new cycles of vegetative growth. A series of ecological changes transformed Porcupine over the years: white pine spread, and the range and extent of the grassland fluctuated, sometimes because of human modification.[9] There is no clear consensus on when and how Anishinaabe people first came to the place now called northern Ontario, but it is likely that several Algonquian-speaking groups and possibly Cree occupied the area that would become the Porcupine mining region when the Europeans arrived.[10] Currently, members of Mattagami, Matachewan, Abitibi, and Flying Post First Nations live and work in and around the mines where they continue to exert sovereignty over that part of the Abitibi and its resources.[11]

To summarize the area's premining settler history very briefly, fur traders had arrived at Porcupine Lake by the middle of the seventeenth century. In the immediate vicinity of the future mine site, tensions with English traders led the French to establish a series of posts in the area, on Nighthawk Lake in 1673 and Lake Timiskaming in 1679.[12] Locals called Nighthawk Lake and the Frederick House River (which flows out of it) Piscoutagamy or Puscoutagamy.[13] After the fall of New France, Scottish American traders competed with the Hudson's Bay Company, creating favorable trading conditions for Indigenous trappers through the late 1700s and early 1800s. Sometimes competition resulted in violence: competing trappers burned the post at Frederick House Lake to the ground in 1813, killing the manager, several laborers, and Indigenous allies.[14] The fur trade caused a steady depletion of furs and big game after 1800 so that by the mid-nineteenth century, many communities had shifted to reliance on fish, small game, and trade goods.[15]

In 1905 Treaty Nine theoretically removed Indigenous title from 210,000 square miles of Ontario (about two-thirds of the province's total area), encompassing the entire drainage basin for Hudson and James Bay. The last of Canada's numbered treaties, Treaty Nine originated from Indigenous requests for support from the headwaters of the Albany,

Missinaibi, and Abitibi Rivers where livelihoods had been most affected by the decline of the fur trade. Indian Affairs and the Department of Crown Lands were happy to oblige, as they saw "Indian title" as a barrier to industry and development. Specifically, proponents argued that the extension of the T&NO railway north of the height of land (and associated resource exploration in the region) required legal cessation.[16] By that time, a 1903 silver rush to Cobalt, lying just within the northern edge of Robinson Treaty lands, had already suggested the possibility of further mineral wealth on the shield.

As part of Treaty Nine, Indian Affairs allocated reserves to individual communities, including for Matachewan, Mattagami, Abitibi, and Flying Post First Nations. Reserves tended to occupy the land around Hudson's Bay Company forts, but the closest fort to Porcupine had been burned in 1813. Porcupine Lake and the Porcupine River were part of a portage route, but other economic centers were relatively distant. By 1909 the state saw no legal or practical barriers to prospectors claiming what they believed was empty land. In reality, Indigenous people lived and worked in the areas claimed by miners but were often erased from historical accounts. Treaty commissioners recorded Indigenous settlements in and around the site of modern-day Timmins during their trip through the area in 1906 and described a winter settlement on Nighthawk Lake maintained for purposes of winter fur trapping.[17] As following chapters show, Anishinaabe people were critical to the initiation of the mining regime, and communities continued to live on and manage the land through the whole mining period.[18]

The numbered treaties relate closely to mining and environmental history in the United States and Canada, and this connection is worth exploring in the context of settler-Indigenous relationships at Porcupine. The first Canadian treaty (the 1850 Robinson Treaty) stemmed directly from land-use conflicts between miners and Indigenous people on Lake Superior during the Mica Bay Incident. In 1845, in response to prospectors' increasing demands for access and regulation, the Crown Lands Department issued regulations for exploring, claiming, and developing land.[19] Local people objected to mining in their territory and wrote letters of complaint to the Department of Indian Affairs. These efforts elicited little response, and Metis and Ojibwa communities attacked the Quebec Mining Company at Mica Bay in November 1849.[20] In 1905 these were the kinds of conflicts the government hoped to avoid by signing Treaty Nine.

The Robinson Treaties set important groundwork for appropriation, including the idea, expressed by Robinson himself, that the land was "barren and sterile" and would "never be settled except in a few locales by mining companies." Treaty Nine failed to deal effectively with usage rights off reserve because commissioners assumed the land was too big and too empty for conflicts over any resource besides (possibly) minerals. Ontario's third Treaty Nine commissioner, Daniel George MacMartin, was described as a mining expert, and his presence on the treaty file suggests that mineral rights were on the government's mind when it drafted the treaty.[21]

The Robinson Treaty and Treaty Nine functioned to alienate Indigenous people from natural resources on their traditional territories. They enabled governments, settlers, and mining companies to own and manage Abitibi landscapes without consent from First Nations people after 1905. Although mining companies are now required by law to consult with First Nations on new projects, most historical development at Porcupine occurred without input from or benefits for the people who had been there first, and consultation continues to be a source of tension between bureaucrats, mining companies, and First Nations to the present.[22]

Throughout this period, the landscape around Porcupine Lake transitioned from a portage route and agricultural and hunting/gathering zone to a location of small-scale mining influenced by international trends in gold extraction. Fire and insect cycles, seasonal Indigenous land use, animal activity, and fluctuating water levels suggest a dynamic ecology even before the miners began to intervene on the land. We can follow these environmental changes and how they were understood in the colonial imagination through the writings of the Ontario Bureau of Mines employees who regularly traveled through the Porcupine and recorded their observations in annual reports.[23] As participants in a broader culture that saw northern wilderness as land waiting to be transformed by a modern industrial regime, bureau surveyors actively sought signs from the landscape as to its destiny. The surveyors described an occupied, managed, and living landscape that nevertheless required colonial intervention to fulfill its productive potential. Their observations overwhelmingly led them to envision the Abitibi as a future agricultural frontier, an alternative destiny that would prove hard to shake.

In 1899 Parks became the first Ontario Bureau of Mines employee to walk over and document the future mine sites.[24] Parks arrived at Porcupine Lake in the spring, so winter runoff rendered the shallow lakes like Nighthawk muddy. He described inlet and outlet streams as tortuously winding and variable, with banks consisting of clay, swamp, timber, and shrubs. Rocky exposures surrounded Porcupine Lake and the Porcupine River more abundantly than other waterways, giving Parks a hint of underlying geological formations (but no indications of precious metals). Parks also noted the scars of recent fires in multiple locations.[25]

In his travels, Parks benefited from the work of fur traders and Indigenous people who maintained canoe routes for their own purposes. Around the Frederick House River, he noted that "several. . .small lakes are scattered over this region, all of which are connected by trails giving evidence of their use as hunting grounds." He described an abundance of otter, fisher, and beaver signs. Near Frederick House Lake, Parks described a small lake with clear water and plentiful fish that "seems to be a fishing ground for the Indians of the region, as many drying places were noticed on its shores." Where routes were not maintained, traveling became difficult due to thick deadfall. "Indians. . .go through this way to the headwaters of the Riviere Blanche, and thence to Lake Temiscaming," he wrote of a series of rapids on his route. "I was also informed that a portage leaves this river somewhere near the first falls and connects with a route to the Matachewan waters."[26] Indeed, until the railway arrived in 1909, all travel through the area around Porcupine Lake occurred on the routes used and maintained by Indigenous people and therefore followed the contours of an economy focused on fishing, trapping, and agriculture rather than mining.

In 1903 a second bureau expedition led by George Kay (and accompanied by agricultural expert Tennyson Jarvis) returned to the ground originally covered by Parks in 1899. In the intervening years, the land had changed. When Kay arrived at the portage between the Mattagami River and Porcupine Lake, he found it in a state of disrepair. Kay also recorded recently flooded landscapes: "Submerged camping grounds and drowned trees over areas of many acres indicated that the season had been one of more than ordinary rainfall." Additionally, 1903 seems to have been a bad year for larch sawflies, which thinned the canopy and left the woods more open to sunlight. Dropped larvae coated the Porcupine River and the shores of Nighthawk Lake, and they were "found in masses" in the vegetation.[27] Originally native to Europe, larch sawflies first arrived in Canada in the late nineteenth century (they were identified first in Quebec in 1882). The species lays its eggs at the base of new shoots, and young larvae strip branches of their foliage, inhibiting growth and weakening healthy trees. In the summer they drop to the ground to form cocoons and transform into pupae that emerge the following spring. Through the early twentieth century, larch sawflies chewed their way across Canada's forests in a series of population explosions and collapses.[28] The party's visit coincided with one such event as it swept through Porcupine.

Aside from the struggling larch, Kay described the area as "wooded with spruce, balsam, birch and poplar; the soil is gravelly sand," "rocky undulating country," with evidence of moose and beaver in local lakes and streams and plenty of pickerel and pike in Porcupine Lake. Mosquitoes and other biting insects significantly shaped their travel. Kay noted that he might have met more "Indians" in the woods, except that "they never hunt during the summer months when flies and mosquitoes are out, but congregate at the forts where they can protect themselves to some extent from the insects by building smudges." Jarvis devoted two full pages to descriptions of flies alone (by comparison, he only used half a page for all other forms of wildlife).[29]

Like Parks, Kay followed existing routes and relied on Indigenous knowledge to fill in gaps of his report. For example, he noted that local people told him "that the Redstone takes its rise in a large muskeg several miles to the west." Kay's visit coincided with the end of the winter hunting season, so he encountered plenty of people during his time in the area. He reported that "Indians were camped on a small island in Night Hawk Lake," and found "potatoes, squashes and onions, which had been planted by the Indians, were in thriving condition" nearby. Kay also noticed huts erected on a bluff overlooking Nighthawk, and he described a place known as McDougall's clearing, which contained a furnished house and garden plus "some Indian Wigwams and a small patch of cleared ground" near Fort Matachewan.[30]

James McMillan led a third bureau expedition into Porcupine in 1905. McMillan expressed more geological imagination than his predecessors: he envisioned the landscape backward to the retreat of the glaciers when it was "a plain, in all probability once the bed of a glacial-dammed lake." When the ice and the water retreated, it left sharp ridges, piles of glacial till, and depressions filled with water that became the area's myriad of lakes. As the land drained, it created the slightly younger *V*-shaped eroded valleys. McMillan speculated that "some depressed tracks" had once been the "beds of shallow lakes. . .now filled with peat and moss to a depth of 4 to 12 feet."[31]

After setting the geological context, McMillan recorded his modern impressions: the larch sawfly attacks described by Kay in 1903 had severely impacted the land. As the weakened larch blew down, it had been replaced by a thick growth of alder: "The killing of the tamarack has left the woods rather open to the sunlight, and a thick growth of alder has sprung up among the more or less scattered spruce."[32] There were other changes: beaver were doing well. In 1899, Parks had reported beaver overhunted, but four years later the animals had clearly rebounded.[33] In 1903, Kay crossed the portage between Jarvis and Porcupine Lakes and noted that the stream was sluggish "due to the fact that beaver are at present operating there, and by their dams are raising the water." Jarvis first traced this activity to human intervention: "Thanks to the wise legislation in the protection of the beaver, this animal is becoming more numerous and the danger of its extermination has been warded off for some time. . . . [F]resh beaver dams were very common."[34] In 1905, McMillan noted on one of the tributaries of the Porcupine River that "a beaver dam 5 feet in height was seen and in fact on nearly every stream beaver cuttings and work are plentiful."[35]

Although the Cobalt silver rush was in full swing a few hundred kilometers to the south in 1905, Porcupine remained a landscape largely shaped and managed by Indigenous people. McMillan's agricultural expert Archibald Henderson took photos of an "Indian hut and potato patch" on the west shore of Frederick House Lake. He also described people who fished extensively, using either trolling or netting depending on conditions. "The larger rivers and lakes of the region are very muddy," he explained, "and for this reason the fishermen must use other means than trolling these waters." Trolling involved dragging a baited line behind a boat and worked poorly if fish could not see the lures through murky water. The solution was nets, which allowed "the Indians" to "catch large numbers of pike, pickerel, whitefish, and, in some localities, sturgeon." In the smaller, clearer lakes and rivers, "pike and pickerel are readily caught with a troll."[36] The latter comment was likely meant to imply Porcupine's potential for sportfishing.

In 1906, H. L. Kerr was the last bureau geologist to visit before the gold rush. Kerr's reports largely follow the same pattern as McMillan's. He started by reaching for the distant past ("at one time a great lake dammed in the north by the retreating ice," resulting in a series of small

Pulled from the 1905 survey of the Abitibi, "Indian Hut and Potato Patch West shore of Frederick House Lake" shows a dwelling and agricultural area cultivated by local people in the area surrounding Porcupine Lake. *Report of the Bureau of Mines, 1905*, vol. 14, pt. 1 (Toronto: L. K. Cameron, 1905), 239.

lakes now filled with muskeg). Indigenous people appear in many of the same places as they did in McMillan's report, particularly on the shores of Nighthawk Lake, and furnished Kerr with supplementary knowledge for his report. Henderson took soil samples from an "Indian Garden" on Lake Kamisokotaia, which he later used as evidence for the region's agricultural viability. Kerr suggested that, even if the beaver were doing well under conservation measures, other furbearers were in decline. Excepting Parks's note on declining beaver populations, explorers had generally reported good game conditions, with plenty of fox, otter, and bear. However, in 1906, Kerr described conversation with "several Indians" who claimed that "game is not so plentiful as it used to be." He later spoke with HBC employees who complained that "furs brought into the fort are becoming less in number each year."[37]

In the end, Kay, Jarvis, McMillan, Henderson, and Kerr all reached the same conclusion. With some encouragement, the North would make good farmland. While they did not exclude the possibility of mineral exploitation, neither were they particularly enthusiastic about it. In Ontario, rich agricultural land had became increasingly scarce in the last half of the twentieth century.[38] Hopeful speculators, boosters, and farmers looked farther afield for free productive land, and their hopes focused on the province's "Great Clay Belt," which promised the possibility for the opening of a new northern frontier. The approval of the T&NO in 1902 stemmed largely from this agricultural vision for the North's future.[39] In addition to subsidizing railways like the T&NO, the Ontario government would build colonization roads, publish pamphlets, reserve homesteads for veterans, establish immigration bureaus, and build experimental farms.[40] As the first step in this process, bureau surveyors had gone into the bush to fulfill the mandate of government to work for the benefit of the public by finding and recording potential economic resources. In their search for "value," the rocks were not an obvious source. Parks wrote that "the rock is barren," but luckily "this region will be a valuable addition to Ontario's agricultural lands."[41] Kerr opined in 1906 that "the chief value of the district is in its agricultural possibilities." Eventually, "railways will connect this part of Ontario with the rest of the country [and] its splendid agricultural possibilities and the various other resources. . . [will]. . . combine to make the future of this part of the Clay Belt particularly promising."[42] In the context of western settlement in the United States and a neighboring settler frontier (at comparable latitudes) on the Canadian prairies, the agricultural dream seemed to make perfect sense. Before 1909 the "promise of reward" was not obviously for the prospector.

Government surveyors recognized that there were inconvenient environmental barriers to be overcome in the pursuit of their northern agricultural hinterland but were confident in their ability to overcome them. Foremost among these issues were muskeg and a short growing season. Muskeg, they argued, could easily be turned to the purposes of the state. The wettest land would "yield a large supply of peat," and the rest "when drained will be available as farming land in the areas in which the soil is suitable."[43] The cold climate was self-resolving: "When the country is cleared, the mean summer temperature will be considerably higher." Evidence was drawn from the prairies: "There is no reason why land, situated in the same latitude as some of the most prosperous parts of the great west, should not some time in the future prove equally productive and support a large population."[44]

There were other possibilities too. Parks reported "immense quantities" of timber available, despite recent fires in the region. He also listed off waterfalls with potential for waterpower.[45] Kerr remarked on "the soil," "forests," "peat beds," "water power," "minerals," and "game" as potential resources. The good soil also meant that the "forest wealth may in time prove to be of no inconsiderable value."[46] The peat beds, although usually seen as a disadvantage, would "eventually prove of enormous value" because of their cheapness as a source of fuel.[47] Waterfalls suitable for power were plentiful, and "we may reasonably expect to see, in the not very distant future, large pulp mills and plants for the preparation of the products of the great peat beds for the market as well as other industries in this northland, in which case these water powers will prove of great value."[48] The only category Kerr was notably pessimistic about was game, which seemed thin. With such abundant potential products suitable for integration into southern industrial economies, northern Ontario's future as a location of *mineral* extraction was far from certain.

Mining Rushes and Early Environmental Change

The discovery of precious metals in northern Ontario changed the character of the imagined northern hinterland only slowly. Aside from Sudbury's nickel boom three hundred kilometers to the south in 1883, the 1903 Cobalt silver rush provided the first real indication of Porcupine's mining future. Cobalt is especially important because Cobalt money, knowledge, and men went on to play central roles at Porcupine.[49] Porcupine was close to Cobalt and the geology familiar. Local prospectors from Cobalt knew what signs to look for, and they knew how to survive in the northern bush. As the annual report of the Bureau of Mines explained, "The geology of the whole region is similar to that of Cobalt. Any person having a good knowledge of the geology of the Cobalt area has little difficulty in mapping other areas in this region."[50] Experience at Cobalt promised that, once found, valuable land would be bought up by capitalists (both Canadian and international).[51]

As metal discoveries multiplied, dialogue around northern development became increasingly focused on preparing the land for an industrial mining future rather than an agricultural one. As mining interests arrived in Abitibi from around the world in response to the news of nickel, then silver, then gold, they brought capital, migration, technology, law, and expectations with them. They formed new ties between Porcupine and an international community of mining zones. Physical environmental change to local geography, especially to local waterways, followed. The colonial government attempted to import the tools and technologies of its extractive contemporaries while avoiding their errors. Results were mixed. The early years of transforming Porcupine into a mining hinterland were marked by negotiation between international forces, the conditions of the local environment, and the many uses to which people were already putting the land. Conflicts between competing visions for the North meant that the first years of development were marked by experimentation and difficulty.

As it had in Sudbury twenty years before, mining at Cobalt began with the railway. As the T&NO pushed into the shield in 1903, construction workers had noticed pink cobalt blooms in the rocks. Both prospectors and geologists knew that cobalt was associated with silver deposits, but few were willing to put in the time and capital in a region seen as "unproven" and "risky" in terms of its mineral potential.[52] Those who eventually staked claims and searched for precious metals in the area depended heavily on the Ontario Department of Mines surveys (like those produced by Parks, Kerr, and McMillan) plus small-scale investors willing to risk a little money for grubstake.[53] Old-timer John Patrick Murphy describes these men as having read "tattered volumes" of the California rush, where they had learned to bite into rocks to test for precious metals among other tricks.[54] Eventually, the samples they produced came to the attention of men with knowledge and means to build a modern mine—Noah Timmins, a shopkeeper at Mattawa, T. W. Gibson of the Ontario Department of Mines, and Willet Miller, a professor of geology at Queen's University.[55] Miller knew that the surface showings indicated deeper deposits, accessible only by (expensive) underground shafts. Timmins agreed to finance a project, and a silver rush was on.

From the beginning, Cobalt was a border-crossing event. Migration, international capital, and imported mining technology shaped both the Cobalt silver rush and by extension the early years of Porcupine prospecting. Popular historian of the northern Ontario mineral rushes Michael Barnes dismisses American involvement in the Cobalt rush as "the old story of a large pool of capital south of the border which knew no international boundary," but American investment in Cobalt was more than just monetary.[56] According to the *New York Times*, Californian placer miners, copper miners from Arizona and Senora, gold diggers from British Columbia and the Yukon, and diamond diggers from Africa all came to Cobalt.[57] There is also evidence that New Zealand expertise had a hand in Cobalt mining development: the Otago School of Mines boasted several graduates working at the Cobalt camp.[58] Cobalt historian Douglas Owen Baldwin notes that "prospectors, writers, stock brokers and mining engineers from New York, London, Brisbane, San Francisco and Johannesburg all journeyed to Cobalt" and describes how scientific presses across Canada, the United States, and England "teemed with discussions" on how best to process Cobalt's complex ores. Until Canada built its own suitable smelters in 1907, Cobalt ores were shipped across the border to mills in New Jersey, Colorado, and Pennsylvania for processing.[59]

In their rush to exploit Cobalt's silver, Cobalt's collection of local and distant engineers, investors, managers, and miners created an environmental disaster. Described by the popular literature as a "town planner's nightmare" and by scholars as an "utterly unplanned community," the town of Cobalt emerged as an afterthought in the spaces between the mines.[60] Streets followed claim boundaries, rather than the contours of the land. The population expanded rapidly: in 1905 Cobalt had more than six thousand residents, by 1911 ten thousand. By 1907 stamp mills were used to break down ore, and "nine such plants graced the skyline close to town." Development was so heavy that it was difficult to access the lake. There were no sewage or sanitary arrangements, and the water became polluted. Eventually, the lake was simply drained so the mining companies could exploit veins running underneath. Cobalt was regularly plagued by smallpox and diphtheria; it boasted the largest typhoid epidemic in Ontario history in 1907 (it lasted six years). Uncontaminated water had to be packed in for drinking. The hasty construction also added to the devastation caused by fires, which destroyed the town in 1906 and 1911. Yet mining interests overwhelmed all other concerns. In its quest for corporate dominance, Nipissing Mine used one of the largest "unskilled" labor forces ever seen in the North (one hundred men) to cut miles of trenches, used hydraulic monitors to jet water over the rock to expose all surface veins, and built an overhead tramway across Cobalt Lake. To provide power for the expanding mines, the Cobalt Hydraulic Power Company dammed Ragged Chutes, Hound Chutes, Fountain Falls, and the Mattabechewan, bringing down power prices by more than two-thirds while significantly altering the function of the watershed.[61]

The environmental changes occurring on the Canadian Shield were well understood beyond Canadian borders. One New Zealand paper told the story of a melancholy sport hunter sitting on the veranda of the Cobalt Hotel: "I spent a holiday shooting through these woods and fishing in these lakes a few years ago," he laments. An interviewer asks, "These woods?" Looking around, "there was scarcely a tree in sight, only a rolling barrenness of cleared land that certainly would not have paid any farmer to clear it, so stony was it all." "Yes," the melancholy man confirms, "but just think of it! If I had scraped the rocks instead of scrambling over them after fur and feathers, I might have hit on a million-dollar vein of pure silver!"[62] In the midst of its own industrial mining boom, the image of stripped land might have felt familiar to New Zealand readers grappling with the arrival of industrial techniques and their subsequent impacts on the land.[63]

As the land around Cobalt got staked up, prospectors fanned outward in search of new deposits. They depended on visual signs to assess the presence of precious metals.[64] Visual assessment remains a primary method of identifying minerals in the field right up to the present, although many prospectors added new scientific techniques to their tool kits after the First World War. The Bureau of Mines knew their audience and recommended prospectors look for "streaky"-appearing rocks. Deposits on the surface would occur in rock that looked oxidized and rusty. Underground, gold-bearing rock had gray or green streaks.[65] Only after a prospector identified ore as valuable based on these visible signs would they send samples to professional assayers for testing.

Early claim staking came about because of Porcupine's place within a broader context of turn-of-the-century extraction, at Cobalt and around the world. Canada, like the rest of North America, subscribed to the free-entry principle whereby prospectors and companies could stake claims on private or Crown land without informing private property owners or Indigenous people.[66] First formalized in the oral regulatory structures of British tin-mining districts and rooted in Roman law, free entry is a manifestation of liberal conceptions of property and ownership. Practically speaking, it ensured prospectors quick legitimacy on the land, despite the presence of Indigenous occupants and even the government's own nascent agricultural dreams.[67] E. O. Taylor of Toronto recorded the first claim in the Porcupine region (just south of the big deposits) on December 20, 1906, and prospectors recorded further claims in July 1907 and 1908 around Nighthawk Lake.[68] One of these later claims belonged to prospector Reuben D'Aigle. A newcomer from New Brunswick, D'Aigle had participated in the Klondike gold rush in 1898 where he had prospected successfully on the Koyukuk River. After working out his claim, he heard news of the Cobalt silver rush. He left for Ontario immediately but arrived too late to join the silver boom. Realizing that he needed a better understanding of geology, he left the North and took a course in minerals at Queen's Mining and Agricultural School in Kingston. At Queen's he encountered the Bureau of Mines reports describing Porcupine. Supported by money from his Klondike finds, he paddled into

Porcupine Lake, staked claims, and did some digging and blasting. D'Aigle's claim did not prove profitable enough, and he left it to lapse before the 1909 prospecting year.[69]

Eventually, the combination of local prospecting with big (international) capital paid off. In 1909 two men from Haileybury named George Bannerman and Tom Geddes found surface gold sufficient to convince them of the region's mineral worth and formed a company funded by Scottish investors to explore it.[70] A party led by Jack Wilson of Haileybury also entered Porcupine in 1909 with the help of an Indigenous guide named Tom Fox.[71] They were backed by Chicago businessmen W. S. Edwards and T. N. Jamieson. The Wilson party found a dome of white quartz, dug trenches around the base, and discovered a large vein of gold. Their find became the Dome Mining Company. When news of the Wilson strike got back to Wilson's hometown, more prospectors headed into the bush, including Benny Hollinger, Alex Gillies, Clary Dixon, and Tom Middleton, who would stake the claims later bought out by the Hollinger Mining Company.[72] Sandy McIntyre and Hans Buttner, a Scot and a German, were right behind and staked two claims on Pearl Lake.[73] They later formed the McIntyre Mining Company.[74] With these three finds, the stage for the next century was set. Dome, Hollinger, and McIntyre mines became the "big three" that dominated Porcupine mining for the rest of the period discussed in this book.

Aside from free-entry principles, Ontario possessed a clear set of rules for staking mineral claims inspired by the mid-nineteenth-century gold rushes and shaped by subsequent developments at home and abroad. By 1909 the law had come full circle from more permissive development-focused regulation inspired by the California and Australia gold rushes to more restrictive laws aimed at capturing some of the returns of mining for the people. In 1909 the first discoveries at Porcupine would be conducted under a centralized system that separated surface from subsurface rights, left default ownership of minerals to the Crown, and imposed royalties.[75] In the immediate aftermath of the rush, the provincial government exhibited a clear understanding of the major issues and policies underpinning extraction around the world and situated the Porcupine within it. For example, in a discussion of gold mining law, the bureau referred to policies made in British Columbia, Nova Scotia, Mexico, and Australia. It chronicled the development of miner's custom from California and argued that Ontario should aim to distinguish itself from the United States. Rather than an influx of miners bringing their rules and customs with them, Ontario needed rational industrial law formed with the benefit of hindsight. Good law would strive to be liberal and encourage free enterprise (a necessity because so much of northern Ontario remained a mystery to the bureau), but should also ensure fair wages and safety.[76] Although "environment" in its modern sense did not feature in this discussion, principles like liberalism, free enterprise, fair wages, and safety all implicate nature—as a valuable product to be regulated or (in the case of safety) as a potential threat to workers.

After news of the first discoveries became widely known in September 1909, "a few thousand claims were staked," according to the Ontario Bureau of Mines. Claim stakers hoped to create a speculation boom in which excitement about the finds would create skyrocketing prices. Although the boom failed to manifest as quickly as most people hoped, some properties changed hands "at fairly large figures."[77] The remoteness of the country and the existence of several mining busts within recent investor memory may have made capitalists nervous or unable to invest in unproven properties. After a slow start, on October 16, 1909, the American voice of the global mining industry, the *Engineering and Mining Journal,* reported on the Porcupine discoveries.[78] By November the journal reported that claims were going for about $1,000 each.[79]

Hyperbolic descriptions helped sell claims and shares to distant investor audiences. Alex Gillies's description of deposits that looked like someone had "dripped a candle along it, but instead of wax it was gold" is featured in the introduction to this book. Later, on another find, Jack Wilson telegraphed his partner to describe a deposit so beautiful it was "beyond description."[80] Wilson was quoted by an Australian paper as saying of the Porcupine deposits that "it is the most wonderful thing I have seen, and I have been in every big gold camp that there is."[81] These accounts were combined with reassuring narratives of human progress. As one visitor wrote, "I have watched with wonder and amazement the progress that has been made within the past week."[82] Ontario was undergoing its predestined transformation to productive hinterland under the hands of hardworking prospectors.

A sense of the changes occurring at Porcupine can be gleaned by comparing descriptions of the land from the premining surveys by Parks, Kerr, and McMillan to post-1909 surveys. Government surveying did not stop just because mining had begun, so the reports provide useful before-and-after snapshots of the region. In 1909 bureau geologist James Bartlett got off the train at mile 228.5 (by the time he arrived, a flag station named "Red Pine Lakes") to record his impressions of the new mining camp. It had been three years since the last survey, and compared to the accounts of his pre-gold-rush predecessors (Kay and company) Bartlett's notes are characterized by tight focus on gold mining. In Bartlett's narrative, the agricultural dream disappeared under the need to support the growth and maintenance of a new extractive industry on the land. The train was the first major difference between his account and the earlier narratives: it had taken Kay weeks of travel to make a journey that took Bartlett a few hours. After getting off the train, the canoe route from the flag station to the new community at Porcupine was about sixty miles, and Bartlett called it "for the most part easy travelling." A "fifty chain" (approximately three-quarters of a mile) portage from the rail line allowed Bartlett to put his boat in the water in a small lake that drained into the Frederick House River and then follow Frederick House River to Nighthawk Lake. On the western shore of Nighthawk Lake, he entered the Porcupine River, which led to Porcupine Lake. The winding nature of the Porcupine River necessitated two short portages overland (three chains and eight chains, or around two hundred and five hundred feet). While his predecessors had moved through the future mining area without pause, Bartlett settled down for eight days at the site with the explicit intention of describing its mineral resources.[83]

Like his prerush predecessors, Bartlett described a low-lying, level plain without obvious signs of hidden wealth. Before the gold rush, this had meant vague speculations about the "promise of reward to the prospector." After 1909 hidden rocks became a major intelligence problem. Bartlett called the area around Porcupine Lake "largely swamp-covered, with occasional outcrops of rock," and as "drift-covered, comparatively little being rock." He talked about the rock as "much altered" (in the geological sense) and observed the presence of quartz veins and quartz schist. In plain terms, this meant that he found the ground difficult to read. Aside from the quartz, there were no recognizable signs that could tell him about the underlying geology. In the South he observed some higher country and more exposed rock, which helped only a little.[84] Bartlett was not alone in his inability to make sense of the land. The *Engineering and Mining Journal* echoed Bartlett's frustrations in 1910, stating that "a good deal of territory is covered with drift, which makes prospecting difficult."[85] Without being able to look at the bare rock, it was hard to know where to dig or how to invest. "Ninety percent of the staking now being done in low-lying country without rock-outcropping cannot show any indications to warrant location under the Mining Act," the journal noted.[86] Over the course of the mining period, the bureau would depend on excavations by mining companies to get a look at the rock, or to provide geological information on their properties conducted by private geologists and engineers.

To counteract the intelligence problem, Bartlett promised a bureau map by July or early August 1910 covering the land and mineralogy of the townships of Tisdale, Whitney, Shaw, and Deloro. Geologists (and the bureau) were under considerable pressure to make sense of the geology quickly so that it could be efficiently exploited. But even where they could get a look under the muskeg, the rock was confusing and complicated because it had been so broken up and rearranged during glaciation.[87] Eventually, bureau geologists turned to similar formations elsewhere in the world as points of comparison. For example, W. G. Millar linked Precambrian rock in Scotland and similar deposits in Ontario. Millar found several points of comparison and touted the usefulness of international examples: "It is of great value to workers here to have an opportunity of visiting areas of rock of like age in other countries, especially where they have been studied and mapped in detail."[88] Others would turn to similar quartz-feldspar and aplite dikes found in the Black Hills, South Dakota; Forty-Mile Creek in Alaska; and research by Louis DeLaunay (who studied Africa, Asia, and other gold-producing nations around the world).[89] When maps did appear, international presses quickly snatched them up. By 1911 the New York–based publication *Davis Handbook to the Porcupine Gold Fields* included detailed maps of the entire region, including lakes, claim boundaries, and the ore bodies of major properties.

One of the biggest changes between Kerr's 1906 report and Bartlett's in 1910 is the absence of Indigenous people. Of course, Indigenous people still lived and worked the land throughout the early years of mining despite their erasure from Bartlett's description. Photographs from the bureau's annual reports show whole families living near the mining zone at Nighthawk Lake. There are also documentary details related to individuals who interacted directly with the industry. Tom Fox occupied the Porcupine Lake area throughout its industrial transformation. Alex Kelso encountered Fox in 1906, clearing a water route with an ax at Frederick House Lake.[90] Fox later served as a guide for Jack Wilson's prospecting party.[91] Nearly two decades after the event, Dome founder William Edwards described "an Indian" who followed a prospector named Colonel Pratt "in his birchbark canoe for three days, had camped with him

The 1911 survey of the Porcupine mining area included this photo of "Indians on Nighthawk Lake," suggesting continuous occupation by indigenous people through the early mining period. *Twentieth Annual Report of the Bureau of Mines, 1911,* vol 20, pt. 2, *The Porcupine Gold Area* (Toronto: Kings Printer, 1911), 9.

every night, and pleaded with the colonel to accompany him to where there was plenty of gold." Possibly this was Fox.[92] Other evidence provides further hints about Fox's ongoing relationship with newcomers in 1909. In a letter, prospector Jack Campbell describes a long list of supplies Wilson sent to Porcupine Lake in the spring of 1909: "two canoes, two sleeping tents and a grub 'cache' tent. . .blankets, cooking utensils, fly nets, axes, shovels, picks, saw, compasses, packsacks, tumplines, [and] provisions." Wilson had Campbell (and three other men) "put the stuff in [to Porcupine] by toboggans and dogs." The men left their supplies in the bush near their claims with a guard "to protect [it] from the old Indian Chief Tom Fox."[93] As a whole, these snatches of information about Fox's interactions with prospectors suggest a man with deep interest in shaping the mining regime. Years after Treaty Nine had supposedly cleared title, Fox still lived and worked on the land and remained near Porcupine Lake until his death in 1926.[94] Fox and other Indigenous people were around in 1909, but Bartlett did not see fit to mention them.

Also differently than his bureau predecessors, Bartlett took no soil samples and provided no comment on timber growth or wildlife.[95] Farming, forestry, and hunting no longer interested him. Although not explicitly environmental, descriptions of development suggest some noticeable changes. There were five claims under development by the newcomers: Robert Bruce, W. H. Reamsbottom, A. E. Way (known as the Bannerman claim), W. H. Davidson, and F. C. Remington (the Wilson property). The Robert Bruce claim had stripped about twenty-five feet of earth from a series of parallel quartz veins that contained visible gold "in grains and leaf-like forms." The Bannerman claim had stripped away earth "at intervals for about three chains [60 meters]." Their trenches were about two feet wide. Davidson had done little stripping, impairing Bartlett's full view of the quartz. The Wilson property was by far the most developed. They had found a new vein only a few days before Bartlett's arrival (the largest yet). Although work had not yet commenced, the vein was visible for 120 meters and was about 40 meters wide. The gold existed here in nuggets "scattered through the quarts [*sic*] in a space of about an inch and a half square to cover a twenty-five-cent piece." In general, Bartlett stated that estimates about the total ore could not yet be made, since "no sinking has been done and very little trenching." A more careful examination would have to wait until "the snow leaves the ground in the spring of 1910." In addition to the mine sites, he observed three embryonic camps on the shores of Porcupine Lake in 1909.[96] The land was changing under a mining regime in noticeable ways.

Destined for development or not, Porcupine would not give up its precious natural wonders easily. The challenge posed by Porcupine's environment became immediately apparent. Wilson's find occurred in the late summer, meaning that staking and shaft sinking occurred in the cold, snowy winter months.[97] Work stalled. As the *Engineering and Mining Journal* lamented in November 1909, "As the majority of the smaller lakes in this section are now frozen over, it is a very difficult matter to get into the country."[98] The weather favored locals used to working through the winter.[99] Once Timmins and the Dome Mining Company pushed roads into the region in 1910, seasonal transport patterns would reverse: easy sleighing made winter the best time for transport because in the spring and summer the muskeg softened.

Compared to the wholesale rearrangement of the environment that would take place later in Porcupine's history, the impacts of early mining were relatively slow and light. As prospectors moved through the bush, they ripped up moss, dug trenches, and otherwise rearranged features of the landscape in search of signs of precious metal. For example, when Gillies and Hollinger were exploring their future claim, Gillies described himself cutting a marking post, while Hollinger pulled up moss to look at the rock beneath. If explorations failed to turn up anything interesting, Porcupine prospectors left tools and trenches in the bush. According to Barnes, "Prospectors commonly found evidence that others had visited an area before them. Such things as adits, or horizontal tunnels, trenches, pits, and even blasting following a vein were usual relics in the gold country." Indeed, when Gillies and Hollinger arrived at D'Aigle's old site, they found his rusted tools there.[100]

Once prospectors confirmed the presence of mineral wealth, they set about stripping the "overburden" (brush, trees, and topsoil) from the land to follow the vein, entailing a physical redistribution of the soil and, if any were nearby, altering local waterways. The Dome property provides the best example of this process. Dome started as a hole in the ground. Miners dug up the "golden staircase" until it petered out at 150 feet.[101] The process involved the removal of soil and its deposition elsewhere, creating holes and piles of rock and topsoil on the land that had not been there before. Without the benefit of industrial equipment, these changes remained relatively contained to the areas immediately around major high-grade strikes.

The scale of processing remained modest in the early years due to the lack of cheap transport for industrial equipment. As the Ontario Bureau of Mines observed, even though "a large number of claims were staked during the fall and winter. . .very little actual mining work has been done in the camp."[102] Early miners nevertheless came up with ways for efficiently processing ore. A flat table of rock found in the bush near Timmins showed signs of being used as an arrastre, a tool originating in the Middle East and used regularly in Mexico and Southern California to grind down gold-bearing ore. Chunks of rock containing gold would be crushed against the table by heavier rocks until the material was fine enough to be run through a sluice or treated chemically.[103]

Yet prospectors could make significant environmental change even with the most rudimentary tools. Father Charles Paradis (prospector and ordained member of the Oblate Congregation of the Roman Catholic Church) accidentally drained the southern tip of the Frederick House Lake in the fall of 1909 to mine the embankment near the lake's exit. Paradis had been digging at the embankment next to the waterfall when the lake began to flow into his excavation. Following Paradis's shallow channel, water flowing out of the lake rapidly widened and eroded its banks, washing away the soil and undermining trees on either side.[104] Assistant geologist C. W. Knight, who explored the area with provincial geologist W. G. Millar, expressed concern about the severity of change on the waterway. Frederick House Lake's water dropped eighteen feet and Night Hawk Lake three feet. The modification left behind "a sand and clay flat with the pinched river winding through," impassable by boat.[105] Knight took photographs of the destruction and compared what he saw to Parks's Bureau of Mines reports of the same waterway. He speculated that Frederick House Lake might eventually disappear because of erosion. This was not necessarily a bad thing: "It may be that future generations of prosaic farmers, long after the Porcupine is a dead one, may bless the worthy friar for draining areas of what might otherwise have remained. . .swamp land." Knight thus expressed hope that today's environmental damage might be tomorrow's boon, and in doing so circled neatly back to pre-1909 notions of the North as agricultural land—a destiny it might still fulfill when mining inevitably finished.[106]

A Rich Man's Venture

As early exploration turned to more sustained forms of development, the character of human interventions on the land changed. Part of this change included the addition of external voices to the documentary record articulating new perceptions and expectations of Porcupine. Taken together, these new voices give the impression of escalating environmental change, including new transportation routes, a rising and diverse population, the formation of small mining companies, the arrival of new technology, and the increasing influence of international capital and expertise. The end result would be a mining frontier that was no longer accessible to individual miners or even small companies, despite clinging to the romantic ideology of the nineteenth-century gold rushes. Instead, Porcupine increasingly depended on a corporate structure and wage labor.

The *Engineering and Mining Journal* indicated that the arrival of big money by the end of 1909 meant changes in the bush: "Several well-known people have lately taken up large interests in the new Porcupine goldfields and have already sent men and supplies to develop their holdings." McArthur and Co. from Scotland had taken over the Way and Griffith claims and planned to sink four shafts to two hundred feet. In January 1910, the journal claimed that "there are about 50 people starting to build stores and stopping places at the northeast end of Porcupine Lake."[107] The assembly of an industrial mining community continued through the winter and into 1910—with some difficulty. After the freeze-up in 1909, Timmins and a crew of twenty-five men got off the train at mile 220 west with two teams of horses and three tons of supplies (including equipment brought in from Cobalt). By January, Timmins's crew had hacked in a road using horse teams to plow through the snow. After a false start on a direct route from the rail line and losing horses to thin ice on Porcupine Lake, he finally arrived at the claim and set his crew to work stripping the vein in preparation for two shafts. Stripping in this context meant removing the overburden covering the paying ore. Cleared lumber became the floors and frames for tents, including an assay office for testing the results of drilling.[108]

Others followed Timmins's lead. M. J. O'Brien of Cobalt shipped in a diamond drill as well as thirty men and supplies to last the winter and entered talks with Alex and Jack Miller.[109] On January 22, 1910, the *Engineering and Mining Journal* ran a full article on the finds and reported that "daily between 50 and 100 men are going in and this number will shortly be largely increased." Citing J. F. Witson, assistant chief of the Provincial Surveys Branch of Ontario, it stated that two thousand claims had been staked, encompassing the entire township of Whitney and Tisdale, "excepting lots reserved for location by veterans, as well as two thirds of Shaw, and two thirds of the township south of Tisdale, which has yet to be named." The journal reported new roads in the district that would run regular stage coaches and asserted that "arrangements have already been made to put steamers on the lake next spring."[110] On January 27, the Ontario Government created a Porcupine Mining Division and appointed Arthur Bruce as mining recorder so that prospectors no longer needed to travel to Haileybury to record claims. The move suggested confidence in the long-term viability of the mining camp and constituted an encouraging sign for investors.[111]

By February the positive trajectory of development at Porcupine finally encouraged some major deals. According to the *Engineering and Mining Journal*, "one of the most important deals yet made in the Porcupine district was consummated on Feb 10 when the Timmins' took up their option on the Hollinger-McMahon properties." The journal excitedly chronicled the acquisition. "[On the McMahon property] it is understood that there is free gold at the bottom. This shaft was sunk off the main vein, but a 4 ft. lead was encountered a few feet down. Sinking will be continued to a depth of 100 ft. And then crosscuts will be run. A plant has been ordered and work will be prosecuted vigorously. The fact of this option being taken up will have a beneficial effect on the new camp."[112] Besides the Hollinger-McMahon deal, the McCormick Brothers of New York purchased the Wilson property for $1.5 million after a two-week investigation by a representative.[113] Wilson's claims would go on to change hands again in 1910 when they were sold to Dome Mining Company for $1.7 million.[114] Money allowed the companies to extend their explorations into expensive underground spaces and build rudimentary processing facilities. Investor confidence and takeovers by large, established firms allowed properties to import expensive supplies and make ambitious new plans for getting the gold to the surface.

In a climate of escalating excitement, and perhaps having learned from the Cobalt disaster, the Ontario government tried to plan for the long-term future of the camp. It made the first steps toward the establishment of a community at the northeast end of Porcupine Lake in February. According to the *Engineering and Mining Journal*, "It has been surveyed into town lots which will shortly be offered for sale." This move was partially to accommodate the "hundreds of prospectors [who] continue to arrive, the total number of people now in the district being estimated at about 4000."[115] As more hopefuls flooded in, they moved into new areas, and the government made efforts to protect parts of the landscape.[116] An order in council protected every lakebed adjoining the Whitney, Tisdale, and Mountjoy townships (or the entirety of the Porcupine district). The government also tried to ensure the right-of-way of the T&NO, which the Ontario Railway and Municipal Board approved in September of that year, creating a Porcupine spur line.[117]

By April the supply road built by Timmins in January (used all winter to bring in people and equipment) began to turn to mud. Before it could dissolve completely, "a few plants were rushed into the camp" so that ore testing could proceed. According to the *Engineering and Mining Journal*, it was "practically impassable for teams, and people coming out or going have to do so on foot." By May the road was entirely impassible. Summer access would have to continue via portage and river from mile 228, or up the Mattagami River, but those routes wouldn't be available until the ice thawed completely.[118] With the road gone, many prospectors could not access their claims or fulfill the development requirements needed to avoid forfeiture. The Ontario government took special measures in light of the situation: "To prevent forfeiture of claims, the Department of Mines has extended the time for the first 30 days' work on claims staked between Jan 1 and March 15 until June 15."[119] As the journal opined, "One of the greatest needs of the new camp is a summer road, and although several have been projected nothing much has been done yet. The government has been asked to assist in this undertaking, but so far has done nothing."[120] Over the summer of 1910, the mining companies moved supplies via wagon as far as they could before transferring them to gasoline boats that traveled fifty-two miles down the Frederick House and Porcupine Rivers to the mine site. Eventually, the mines developed a seasonal rhythm for transportation shaped by the realities of northern Ontario muskeg. Supplies were moved in the winter, "since roads which are impassable in the summer are excellent in the winter when used for sleighs."[121]

The combination of massive capital requirements with an inaccessible and difficult landscape understandably made investors nervous. Since the Porcupine was remote and difficult to access, distant investors were forced to put significant trust in the men on the ground. Frank Cochrane, Ontario minister of lands, mines, and forests, sounded the first note of warning in January 1910. He stated that information had reached the department that "'snow-shoe staking' is going on extensively in the district, and that it is only fair to warn the public against buying claims staked out when the whole country is covered deeply with snow, as there can be no certainty of the *bona fide* discovery of minerals under such circumstances."[122] His concerns about snowshoe staking proved well founded. Later that year, two Porcupine prospectors, E. E. Jones and A. Blackburn, were charged with causing a rush to Camel's Back Lake. According to the *Engineering and Mining Journal*, "About a thousand gold-seekers were deceived by erroneous reports," and "they are alleged to have sold claims to Haileybury men for $2000, signing affidavits as to their value."[123] Distant investors depended on honesty and expertise from prospectors and geologists about the nature of gold deposits and their value—a common state of affairs in the world's emerging industrial mining frontiers exacerbated by Porcupine's relative inaccessibility in 1910.

For individual prospectors and small companies without capital, the cost of admission to Porcupine's deposits had rapidly escalated well beyond reach. Boosters sold the Porcupine as the successor of Sutter's Mill, Victoria, and the Klondike with the same rhetoric of quick riches for enterprising men. When hopeful immigrants arrived at Porcupine, they quickly found themselves shunted into mining labor for growing companies. In 1910 a Scottish immigrant (who wrote anonymously) recounted his experience of the Porcupine rush. After arriving in Canada, he hopped on the T&NO in Toronto toward Porcupine. He "was dumped at midnight along with about 500 others at what is called '222'" (the railway mileage), at a small town "where but three weeks before there was nothing but the virgin bush." In the morning he "woke early and. . .had a wash in melted ice, then a rush through the bitter cold to the cookhouse [for a] hearty breakfast of porridge and hash." Anticipating finding wealth in Porcupine, he took the expensive stagecoach instead of walking to the gold camp (which would have taken him three days). The journey was cold and rough and required that he occasionally jump off the coach to warm his numb limbs. The stage stopped at noon at Father Paradis's halfway house (on the shores of drained Frederick House Lake) where he ate again while the horses rested: "Food doesn't stay long with a fellow in that country. You seem to breathe a mouthful out with every breath."[124]

Once arrived at Porcupine, the man described Main Street as "uneven rows of log shacks." He found lodging and began to look for work. It took him three days to find a job, which ultimately disappointed him.

> To tell of the bitter cold and hard graft, and the tugging and pulling of four and five hundred weights of provisions and tools on toboggans, the shovelling of snow, the stripping of rock, and the sinking of a shaft would take too long. . . . I put in about thirty-eight days at 2½ dollars a day and board. The man without capital has small chance here, as capitalists and prospectors have staked every inch of ground within a radius of thirty miles. It is a rich man's venture.[125]

By the end of 1910, large capital-intensive companies dominated Porcupine. The nature of the deposits gave rise to a mining culture that promised the get-rich-quick opportunities of the nineteenth-century gold rushes in ideal while depending on organized international capital, labor, and technology in practice.

This individual experience of Porcupine is instructive in that it suggests that industrial transformation excluded individual and small-scale mining from an early date. For hired laborers, work could be difficult and dangerous.[126] The story also shows that the northern climate permeated human interactions of the land. The snow and cold stand out as definitive parts of life, shaping movement, work, and mental states. Other northern mining frontiers had produced similar experiences. In the Klondike, firsthand accounts describe boatloads of "disgusted people getting out," scarce work, and disappointing wages.[127] Yet the account also documents work that was primarily outdoors, aboveground, and varied. The conditions of work would stabilize somewhat at Porcupine by the end of the First World War, as the labor needs of the industrial giants created clear categories of skilled and unskilled work. Nevertheless, the instability of mine work, miner transience, and opportunism continued to characterize the work of extraction into the mid-twentieth-century mining fields of northern Canada.[128]

Although some hopeful miners came to Porcupine in search of wealth, most people who interacted with (and profited from) the Porcupine did so from afar. In a world that had learned to appreciate spectacular stories of frontier bonanzas, the Porcupine gold rush obtained easy fame despite its remoteness. International media watched events unfold in Canada with speculative interest based in experience garnered from California to the Klondike.[129] The Abitibi had become part of a mining continuum spread across multiple continents. Porcupine stories fulfilled specific narrative functions for a broad audience with shared expectations of gold mining. Specifically, wild stories of intense hardship invented by the international press reflect pessimism about the nature and risk of mining on the frontier in the early years.

Enthusiasm for the Porcupine discoveries was muted somewhat by the fact that, by 1909, people had already heard the cry of "gold!" sounded prematurely by speculators and "wildcatters" so often that their first response was skepticism. As the *Otago Daily Times* put it in December 1909, "The people of Ontario are much worked up over the discovery of new and promising goldfields in the northern part of the province." This excitement was partly a product of the success of Cobalt, but "account seems not to have been taken of the fact that many more millions will have to be taken out of Cobalt before the profits equal the actual outlay." Furthermore, at the Porcupine, "the gold is in quartz, and that means large capital before any return can be had." In another article, a journalist wrote that "so far, a large number of prospectors have, of course, found nothing but disappointment." In 1911 the *Goulburn Evening Penny* reprinted a story by the *Canadian Courier* that was "sarcastic regarding the latest gold discovery in the Dominion at Porcupine, in Northern Ontario." Apparently, "two real gold mines have been discovered, and 8000 mining claims have been staked. The other 7998 are still prospects." This meant that "the chances of the man who invests in Porcupine companies is about one in 1987. The small investor who takes that chance must either be a gambler or a fool." The *Penny* observed that "Canada has plenty of experience to guide her citizens in relation to mining speculation. Probably twenty-five million dollars was lost in the Rossland boom. The mining brokers and the incorporators of mining properties got most of that." The *New York Times* reported on the fact that the "Curb Association" needed to keep "wildcat stocks" out of the booming market for Porcupine stocks. The Australian *Northern Miner* ended an otherwise bland description of operations on the new Porcupine goldfields with this statement: "Nevertheless, the camp will yield its full quota of wildcats, and many speculators will be badly bitten."[130]

The international media sensed that success or failure at Porcupine would touch distant local economies in important ways. The New Zealand paper the *Star* worried that the silver from Cobalt would drive down the price of silver. "Thousands of Canadians and Americans are suffering from the Cobalt fever—not the typhoid, but something rather worse, the share-gambling mania. Shares in the paying mines stand high; but many companies have failed as yet to discover anything but an easy way of sinking shareholders' money in holes in the ground." In Kalgoorlie, Australia, the *Sun* urged investors in its mining investment section to ignore the hype around Porcupine and instead spend investment dollars at home, on Australian goldfields in need of exploration and development.[131]

Skepticism and worry could also be seen at home. "The story sounds like similar stories that have started or rather added to other stampedes into gold camps," the *Globe* (Toronto) pointed out in December of that year.[132] The newspaper compared Porcupine to an earlier failed rush at Larder Lake: "The early history of Larder Lake country was such that it disappointed the public."[133] Even though gold existed at Larder, a lack of proper equipment meant that the "rush" quickly petered out. A gold rush was not a guarantee of profitability, and the *Globe* was not yet convinced of Porcupine's viability. The paper took the indifference of experienced mining men as a sign that the finds were not worth getting worked up about yet: "It is a notable fact that right in the Cobalt camp there is not much enthusiasm just yet." "The Cobalt camp" consisted of a number of experienced prospectors, miners, and businessmen, many of whom had been educated abroad. Besides, the *Globe* was wise to the fact that even good showings of gold would mean that "there will inevitably be hundreds of [prospectors] doomed to despair" and only a few who would actually profit.[134] In addition, alluvial or surface gold had become a signifier of deeper wealth rather than an end unto itself. Readers of the *Globe* knew that the presence of quartz meant that Porcupine's gold would require mills and men with capital to extract. Porcupine was not rich in surface gold to begin with—although "one optimistic prospector alleges that gold to be the value of $7,000,000 is visible on the surface of the field," readers knew not to take this sort of speculation seriously. "Official channels," the *Globe* reported, called the report "decidedly enthusiastic," and readers waited for government reports of the field's real value.[135]

Whether international or local, interest seems to have been largely restricted to the investment possibilities of the new rush. The likelihood of newspaper readers packing their bags and heading off to make their fortunes had diminished significantly since California in 1848. For example, in January 1911 the New Zealand *Press* reported that "English and Canadian capital is flowing in for the purchase of claims in the new goldfield at Porcupine Lake, on the borders of Hudson Bay." In the *New York Times,* ads ran for market letters for interested investors. The *San Francisco Call* also ran advertisements for investors.[136] Readers were still expected to participate in the Porcupine gold rush, but their involvement would remain less direct than it had in the 1850s and 1860s. Where nineteenth-century gold-rush media coverage had often included maps, information about purchasing supplies, and tips for prospectors, the Porcupine coverage focused on how interested investors might direct their capital most profitably.

Aside from a possible investment opportunity, Porcupine made for a sensational story of frontier challenges. By 1909 gold rushes already had an associated mythology for literate news consumers. The gist of this mythology involved adventurous men in adverse circumstances struggling against an untamed wilderness. Starvation narratives were one way in which this manifested. The dramatic and well-publicized famines of the Klondike gold rush may have been fresh in the minds of news readers—only a decade earlier, thousands of hopefuls arrived at the Chilkoot pass unprepared for the fact that there were no supplies of food, resulting in food shortages in the city of Dawson and Skagway in 1897. There were also stereotypical stories of bush hardship, gambling, and murder that made international news about Porcupine. In 1910 the *Woodville Examiner* told the story of a prospector who shot his partner over a claim dispute.[137]

Reports of starving prospectors, greed, and violence made for good headlines, even if they were not particularly accurate. This resulted in a short bout of false reports about the Porcupine. In November 1910, newspapers in New Zealand began reporting that Porcupine prospectors were starving. On the seventeenth, the *Hawera & Normandy Star* reported, "Scores of prospectors are dead and others are dying on the trails to the Porcupine gold camps, in Northern Ontario, through the food supplies failing to get in." The next day the New Zealand *Colonist* replicated the report verbatim. The same story was repeated by the *Akaroa Mail and Banks Peninsula Advisor*, the *Wanganui Herald*, the *New Zealand Herald*, the *Grey River Argus*, and the *Otago Daily Times* in New Zealand.[138]

Juicy details added legitimacy to the story. In Australia a story titled "Awful Suffering: Early California Again" repeated a report by "Mr. Mclean, who has had widespread experience in Australia, South Africa, and Alaska," that "scores of prospectors were either dead or dying on the trails to the Porcupine gold camps in Northern Ontario. He considers that the suffering being experienced surpasses that on the early Klondyke fields, and equals that in California when many hungry men ate their dead comrades." The report went on:

> Many who sought the new rush had meagre supplies with them, while those having plenty abandoned them. The impassable roads, and the barren and desolate country soon produced a famine. . . . [T]wenty of the unfortunates are already dead, and hundreds are dying. Snow has obliterated the trail, and owning to this two men became engulfed in a quagmire.[139]

The story was repeated in different versions in numerous Australian papers.[140] Some papers reported that starvation was actually in the Alaskan Porcupine camp, reflecting the lack of familiarity with Canadian geography.[141] The false stories were later retracted by most papers, who admitted that "it is officially reported that no deaths have occurred in connection with the Porcupine rush, though great hardships have been endured and many are starving."[142]

By 1911 the story had morphed to focus on the unknown dangers of Canada's mysterious northern landscape. According to the *World's News* in Australia, at least five men died after sinking into the mud around Frederick House Lake. The story titled "Engulfed in Mud: Gold Seekers Sink in Presence of Onlookers" told the tale of miners "swallowed up by the 'honey-pots' that lie around Frederick House Lake." The story apparently came from W. R. Macleay, a mining engineer, "who, emaciated and gaunt visage, has come back to civilization [Montreal] recover his health." Apparently, autumn rains cut off supplies. "Nearly everything edible had been eaten, and starvation stared the miners in the face. They were therefore confronted with a slow and miserable death, if they remained in camp, and until the frosts hardened the surface of the swamps they would encounter a no less terrible fate in striving to reach the outside world unless acquainted with the trails." Leaving camp proved no less risky: The most difficult and deadly part of the journey was along the shores of Frederick House Lake, at the site of the old HBC post. "It was on this part of the trail," said Macleay, "that I saw. . . two men with loads on their backs slowly sink down in the mud and disappear from sight. Their cries and shrieks were frightful to hear. . . . [T]he sight and sounds were horrifying." Macleay went on to talk about how "sticks made out of boughs of the trees were pushed down to probe the depth of the mire, and it was a legend in many parts you might go down 20 ft or more before solid ground was touched when the rains came in such quantity as had then fallen." Macleay told of three other similar deaths he had heard of as well.[143]

In general, international coverage of early Porcupine mining fell into established genres of gold-rush storytelling that had reached their apogee in previous rushes to California, Australia, New Zealand, and the Klondike. International stories also reflect the fact that, by 1909, literate audiences associated gold rushes (and northern landscapes generally) with certain historical tropes; namely, the idea of adventurous men enduring unimaginable hardships in desolate, mysterious northern landscapes encouraged hyperbolic and occasionally false stories of starvation and death. In his study of British coverage of the Klondike gold rush, geographer John Davis lists weather, travel, costs, and starvation among the topics most likely to be covered. These subjects apparently remained interesting in 1909.[144] Unlike the Klondike, however, Porcupine coverage reflects the normalization of the idea that foreign investment rather than individual effort drove the mining industry a decade later and in a place where alluvial gold proved scarce. Because of experiences with the rampant speculation that characterized earlier gold rushes, the climate of Porcupine investment was overwhelmingly distrustful. Unlike on earlier mining fields, where newspaper reports of gold finds played instigating roles in the subsequent rush, readers in 1909 did not plan to go to Porcupine themselves. They were interested in developments for more general cultural or economic purposes. These new purposes for the Abitibi shaped the way it was portrayed (and frequently misportrayed) by international presses.

Geology, History, and the Making of a Mining Camp

By 1909 the discovery of gold implied a particular pattern of human actions that had been developed and practiced on goldfields around the world. Surface indicators showed prospectors where buried mineral wealth lay under a thick layer of overburden. An established investment community stood ready to provide capital to proven goldfields. The nature of Porcupine's ore bodies, which existed in hard rock, discouraged individual prospecting and encouraged the rapid transition to capital intensive industrial extraction based on models developed in California, Australia, British Columbia, New Zealand, South Africa, and the Klondike. Granted legitimacy by the state and energy by enterprising individuals (settler and Indigenous), mining transformed the land. By the end of 1910, extraction began organizing under large corporations particularly well suited to the capital-intensive work of getting gold out of the Canadian Shield. A set of power relations in which individual miners worked under the auspices of international investors and corporate owners to wring profit from the northern landscape prevailed.

The arrival of the mining industry in 1909 marked a moment of human and environmental change at Porcupine related closely to existing mining regimes elsewhere in the world. In a special report on the district published early in 1911, the Ontario Bureau of Mines accurately stated that the Porcupine discoveries represented a new era in history in which the province of Ontario would become a global gold producer.[145] That year American author Harold Palmer Davis produced *The Davis Handbook of the Porcupine Gold District* in New York, a reference guide for interested international experts and investors. The book included maps and descriptions of the landscape that functioned to render the remote North legible to distant readers. Its list of promising companies showed the true internationalism of Porcupine: a combination of South African and American interests controlled the Rea Mines, Americans owned the Dome, an English group ran the Northern Ontario Exploration Company, and most of the other major companies enjoyed similarly worldly connections with only a few truly Canadian-based enterprises.[146] The rapid environmental change at Porcupine after 1909 rested on a foundation of border-crossing capital and experience empowered and encouraged by the Canadian state.

Yet from the perspective of the development-driven Bureau of Mines and the mining companies' executives, much work remained to be done. In comparison to other more mature mining zones, the

Porcupine remained primitive. The pervasive cold, muskeg, and isolation of the camp limited mining development by preventing dissemination of investment knowledge and restricting the amount of equipment that could be shipped in. The tensions between expectations for the new gold zone and practical environmental barriers continued to haunt Porcupine over the next twenty years. The existence (and persistence) of Indigenous economies and the emphasis of early surveys on agriculture meant that extraction never enjoyed total dominance over the land.

The discovery of gold at Porcupine Lake occurred within overlapping historic contexts. Events at Porcupine were driven by the Province of Ontario's conviction that the North would serve as a profitable economic hinterland. This drive for development directly influenced the treaty-making process meant to remove Indigenous title to Porcupine and, later, the decision to push a railway into the Porcupine region. At the same time, mining communities around the globe had already done much of the work of establishing a tradition of neo-European relationships with gold-bearing ground. An infrastructure of scientific and investment ability, technology, and knowledge existed ready-made for Porcupine and was practiced at Cobalt and then rapidly applied by men like Jack Wilson and Noah Timmins farther north. Yet the existence of gold in shield rock around Porcupine Lake remains only a single characteristic of a dynamic, complicated, and variable landscape. The gold camp that had emerged by 1910 represents the point of overlap between these local and external historic forces.

CHAPTER TWO

The Great Fire

Clearing the Way for Economies of Scale After 1911

International presses may have exaggerated tall tales of death in an unforgiving northern Canadian wilderness frontier in 1909, but in 1911 far-fetched stories became terrible reality. On July 10, high winds turned smoldering brush fires into a massive wildfire that moved rapidly toward the Porcupine goldfields. Residents scrambled to load wagons with belongings and headed to shelter in lakes and mine shafts. Not everyone found safety: people and animals burned in the flames, and others drowned in the storm-ravaged water or suffocated in the shafts. By the time the cataclysm was over, it had claimed seventy-three lives, burned 553,000 acres, and caused $3 million in property damage.[1] At the time it was called the worst catastrophe in Ontario's history.[2]

Regular fire cycles are a normal part of Canadian Shield ecology. The annual reports of the Bureau of Mines recorded many large burns in the area around Porcupine the years before the discovery of gold. So why was 1911 so catastrophic? The province of Ontario began actively suppressing fire following the 1897–99 Royal Commission on Forestry Protection in Ontario. In general, suppression meant that rangers suppressed low-intensity fires that might otherwise have burned up local fuels to protect marketable timber, settlements, and infrastructure. Although ranger coverage remained patchy in the north to 1909, suppression by forestry operations may have exacerbated natural fuel accumulation from insect-kill and blowdowns. Logging operations associated with the mining boom and the recent accessibility of local forestry blocks created additional fuels in the form of piles of flammable slash in the bush.[3] Then North America entered a cycle of warm years around the middle of the first decade of the twentieth century. Numerous small fires burned around Porcupine and were suppressed in 1910 and early 1911, foreshadowing the July blowup. The fire on July 10 escaped control when human and natural forces worked together to create a perfect storm.

The Great Fire exposed fault lines in the relationship between humans and nature at Porcupine. This relationship had begun when miners arrived in 1909 and formed syndicated companies. In the three years since gold's discovery, miners, companies, and governments had painstakingly assembled networks of labor, family, technology, science, experience, legislation, investment dollars, and ideology. Some of the features of this new international-industrial regime proved incompatible with existing ecology. Fire patterns required adaptation if the mines were to be viable in the long term, so the newcomers aimed to fireproof Porcupine after 1911. In the hands of the mining industry and its promoters, reconstruction in its aftermath symbolized transformation from wild and dangerous frontier environment into modern, massive, and well-connected resource hinterland.

Chapter 2 argues that the Great Fire of 1911 literally and metaphorically cleared the way for the dominance of large-scale industrial vision for mining based on international examples—with attendant consequences for working conditions. It rearranged physical geography in ways that benefited some stakeholders more than others. Well-financed mines like Hollinger, Dome, and McIntyre bounced back quickly. With their insurance settlements, they built new fireproof infrastructure, studied the newly exposed shield rock using international expertise, and bought up the properties of their burned-out neighbors. By the outbreak of the First World War, the "big three" mined on an unprecedented scale. Environmental problems during and after the Great Fire of 1911 are tangled up with the acceleration of large-scale mining, growing inequality between people, and increased integration between Porcupine and its international context.

Fire Ecology in Northern Ontario

Ecological research on fire cycles shows that fires completely burn Canadian Shield forest in approximately hundred-year intervals. Within that hundred-year cycle, warm dry periods occur approximately every ten years and are marked by larger and more severe fires that burn larger areas than fires in cooler, wetter years.[4] Certain factors can further increase the likelihood of major or catastrophic burns within the cycle. For example, insect infestations and trees downed by wind can increase the amount of dry fuel and contribute to especially big fires within normal ecological cycles. Before colonialism, First Nations people played a role in fire patterns by burning forest for clearing, cultivation, and hunting.[5] The year of the Great Fire of 1911 was within a natural cycle of hot, dry weather in Ontario, but human factors increased its scale, intensity, and severity.

Reports from the Bureau of Mines and climatic data from the early twentieth century provide a record of this relationship.

Insect infestation, specifically European-imported larch sawfly, killed many of the trees around Porcupine at the beginning of the twentieth century. In the 1904 annual report of the Bureau of Mines, Tennyson Jarvis recorded the state of the larch sawfly infestation in the forest along the Porcupine River. "Nearly all of the larch or tamarac trees in this northern country have been destroyed by the larvae of this saw fly," he observed. "During the early part of July the adult flies were seen floating down the Porcupine River, and a few days later the shore of Night hawk Lake was covered with them. Pupa-cases were found in masses beneath the surface of vegetation of all the trees examined in the district."[6] Later reports echoed Jarvis, recording large numbers of trees dead because of the larch sawfly.[7] The infestation resulted in considerable fuel on the ground before miners even arrived in the area five years later.

In 1910 La Niña winds brought down additional windfall across North America, and historians believe that the 1910 winds contributed to a series of massive, deadly wildfires in the United States.[8] The annual reports before 1911 record considerable windfall in Ontario.[9] In nearby Prosser Township, McMillan wrote in 1905 about the "many fallen trees in the Clay Belt" and called it an "extreme condition." "In this area there are very few standing trees. In one locality there is a stretch of twenty chains without any. Spruce, poplar and tamarack lay piled over one another, so that for chains at a time one can walk over tree trunks without touching the ground at all."[10] Surveyors were interested in windfall largely because it represented wasted timber and lost value. Larch-fly attacks may have weakened stands and contributed to high numbers of downed trees.

Fire ecologists now know that state intervention in fire cycles associated with colonial settlement intensifies the length and severity of wildfires.[11] Fire suppression began in Ontario in 1886 with the appointment of the first forest fire rangers. The fire service slowly expanded its mandate and increased in effectiveness over time. Ontario fire policy, like that in the United States, emphasized total suppression. This mandate remained part of Ontario fire policy until the late twentieth century.[12] Documentary evidence does not show whether fire rangers worked around Porcupine between 1909 and 1911; the earliest newspaper account recording their presence comes from 1915.[13] However, the arrival of the rail line and the presence of timber interests would have made Porcupine a good candidate for ranger surveillance, and miners worked to suppress fires even when there were no rangers around.[14] Although mining disturbed the earth and cleared timber, it also left piles of waste wood—mainly branches and treetops unsuitable for shaft timbering or other building and covered in flammable needles.[15] Plenty of fuel capable of sparking catastrophic fire lay between and around the mine sites by 1911.

The impact of a period of hot, dry weather around the turn of the century is visible in Ontario's fire record. Forest historians record large fires extending from Kabinagami to Little Abitibi Lake, Lake Kesagami, and Grand Lake in 1901, just north of the future mining zone.[16] Most post-1900 reports from the Porcupine mention burned patches in and around Whitney and Tisdale townships.[17] In 1905 the annual report lamented the fact that "most parts of Northern Ontario. . .have been swept of their timber by forest fires which devastate parts of the country almost every year."[18] In 1907, W. L. Goodwin struggled to conduct mining classes because of interruptions caused by fire. When he arrived at the Orange Hall in Haileybury to give a lecture, he found he could not use his electric lamp because "the sawmill with which the power house was connected burned down." A few days later while examining an ore pile at the Benn mine, Goodwin "noticed a very heavy explosion towards the east." He later found out that "bush fires, which were burning everywhere in the district," had blown up a dynamite magazine. Just a few days later, at Giroux Lake, Goodwin arrived at University Mine for a class and instead found "the miners fighting a fire which was threatening the buildings of the Foster mine." Rather than teach, Goodwin spent the day fighting the fire alongside the miners.[19]

Fire historian Stephen Pyne lists this early-twentieth-century warming as one of the major contributing factors to the devastating forest fires in America in 1910. The Great Fire of 1911 was only one of many similar blazes at the end of the first decade of the twentieth century—the consequence of coincidentally warm dry years on multiple continents. Between 1909 and 1914, forest fires in Siberia were "as common as blackflies."[20] Australia experienced serious fires along the coast in 1910.[21] Of particular interest to Pyne, catastrophic fires raged across the western United States in 1910 and 1911.[22] Although unconnected to these other blazes from an ecological perspective, the Porcupine fire fits into this broader cultural context of fire experiences on the margins of European and neo-European states. Foreign newspapers sometimes grouped news of the Porcupine Fire with accounts of other fires that summer, especially those in North America.[23] The Great Fire occurred at the intersection of global and local fire trends stretching across both time and space. Not only was it part of the natural reoccurring cycles stretching back to the beginning of shield forest ecology but it was also part of a wider experience of fire in 1910 and 1911.

Ontario experienced some fire in 1910, but conditions worsened in 1911.[24] Data from Ottawa (the nearest city for which records exist) suggests that temperatures were even hotter and dryer in 1911 than 1910. Ottawa recorded its hottest ever days on July 3, July 9, and July 10 (the day of the fire) in 1911.[25] In the period between 1880 and 2016, 1911 has the fourth-highest daily temperature ever recorded (36.7 degrees Celsius or 98 degrees Fahrenheit). Meteorological records for Ottawa show that 1910 and 1911 experienced well-below-average rainfall and low snowpacks.[26] Anecdotal evidence supplements the data. The *Globe* recorded fifty-eight heat-related deaths and dangerously low reservoirs in Toronto on July 8, 1911.[27] Dominion horticulturalist W. T. Macoun recorded a hot, dry spring for Ontario generally (which shortened the flowering season for many blooms) and noted that "July was an extraordinarily hot month, one of the hottest ever experienced." According to Macoun, the mean temperature in July was a scorching 97.8 degrees Fahrenheit (36.5 degrees Celsius), nights remained hot, and rainfall was light.[28]

Although small fires surrounded Porcupine in the early twentieth century, the immediate site of the mines had not experienced a major burn since the beginning of the written record. Just a few months before the Great Fire, the Bureau of Mines observed that "much of the timber on the higher ridges was burned by forest fires. In this district, however, the forest fires are not wide spread."[29] Minor blazes, such as one that leveled Hollinger's almost-completed thirty-stamp mill on May 19, 1911, were quickly extinguished.[30] Fuel on the ground had never burned off as it had elsewhere. Insect kill, windfall, hot and dry weather, settlement, railway completion, and fire suppression combined to create perfect conditions for a destructive wildfire at Porcupine in 1911.

The Great Fire of July 10, 1911

On July 10, 1911, several small fires burned adjacent to the mines and communities at Porcupine. In the morning, the wind united these small fires into a single, much larger fire that began moving toward the town. When the danger to structures and people became evident, women and children began crossing Porcupine Lake to Golden City. Bad weather made the water choppy, so progress was slow.[31] Meanwhile, men and women remaining in town tried to protect it with bucket brigades.[32] As structures in South Porcupine began to burn, people waded into the lake with their animals and belongings, using pieces of wood and other floating material as rafts. According to local historian Michael Barnes, the last telegraph came out of Porcupine at 3:30 in the afternoon to report that the fire "possessed" the town and the smoke was too thick to see through.[33] Fire followed smoke, consuming the town and many of the nearby mines over the course of the afternoon.

The first deaths occurred in the outskirts of Porcupine, where prospectors worked in the bush and could not get out of the fire's path. Within the town, some did not make it to the water or died in burning buildings while trying to rescue relatives, animals, and possessions. Others drowned in the rough lake water or sought inadequate shelter in small streams or mine shafts. Robert Weiss, manager of West Dome Mines, took his wife, child, and twenty-two workers into the Dome shaft where they all died of asphyxiation when the fire sucked the oxygen out. Even after the fire was over, others died from shock or injury.[34]

In addition to human and animal lives, the fire destroyed nearly all the infrastructure built at Porcupine since 1909. The Porcupine extension from the T&NO railway, mine headframes, homes, and outbuildings all went up. Photographs show a smoldering landscape dotted with the twisted remains of metal equipment.[35] J. B. MacDougall, a local teacher and author who claims to have arrived on the scene the day after the fire, wrote that "for those of us who knew the land before, no words can tell. A stark, black, smoking shamble as far as human eye could reach—a very charnel-house of death and desolation."[36] Porcupine newcomers had celebrated the frontier spirit of development that had transformed northern Ontario into an industrial gold center, but the 1911 fire reduced these signs of progress to ashes and rubble.

The fire abruptly severed Porcupine's connection to the world. Paradoxically, the abrupt severing of Porcupine's cross-border ties made them more visible in the documentary evidence. International migration of people, ideas, and objects between nineteenth- and early-twentieth-century goldfields is tricky to follow because census records in frontier zones are frequently incomplete. In Porcupine private and informal human connections between mining zones became matters of public concern and government intervention in the aftermath of the fire. The international media immediately picked up and disseminated the story, suggesting the extent to which international audiences had a stake in

Photograph of Porcupine after the 1911 fire shows a smoldering landscape dotted with the twisted remains of buildings and equipment. "Remains of bank after fire," CPC-00885, the William Ready Division of Archives and Research Collections, McMaster University Library.

Porcupine affairs. Worried friends and relatives of people living at Porcupine unable to contact their loved ones began to channel requests for information through the Canadian government. As a result, personal correspondence was archived within government fonds. These records give a sense of the nature of Porcupine's largely informal international connections as they had been built since 1909, especially when combined with media reports.

A selection of press accounts suggests that the story spread widely and rapidly in the days following July 10 both in the popular media and in mining-specific publications. The wide dissemination of these stories suggests Porcupine's cachet with overseas audiences and the specific ways Porcupine mattered. On July 14, 1911, the *New York Times* ran a short piece noting that, despite the news of the fire, share prices for Porcupine mining companies had not dropped significantly.[37] However after this rather bland first report, the *Times* fell into the trope of fantastic tales of frontier danger. Two days later, it published an account of a mining engineer supposedly just returned from the scene. The account was exaggerated. It claimed "thousands, perhaps, dead, women and children were scattered all over the field of havoc," and described massive losses of life and property.[38] The *Times* published a third account on July 18 from another survivor who had been living in a tent near Pearl Lake and had only enough time to stuff a small pack before running into the water. He stayed in the water with a fellow prospector until it was over and recalls sharing the lake with a bear and other animals escaping the flames. After it was over, the two prospectors hiked out of the burned zone to Golden City and found the bodies of seven men on the way out.[39] Such accounts of dangerous frontier life may have served to fulfill cultural tropes about the character of miners and gold-rush violence. The *Wall Street Journal* also covered the fire. Even though it was only days after the blaze had ended, the paper emphasized that the damage had been exaggerated; Dome Mines could easily have its mill going again by December.[40] This was probably aimed at reassuring American investors who had learned about the fire from other sources. In fact, the fire totally destroyed the Dome mill, and a new mill did not replace it until March 1912.[41] The difference in coverage between the *New York Times* and the *Wall Street Journal* illustrates the two main ways that Porcupine mattered to readers outside Canada—as an exciting story about the hazards of wilderness frontiers and as an event with potential influence on global investments.

Interest extended beyond North America. By July 14 and 15, major papers in Australia and New Zealand reported more than three hundred deaths and "many millions of pounds" in damage.[42] In addition to the familiar sensationalism about the danger of the Canadian wilderness, the Australian papers were especially interested in circulating stories about the "cowardly behavior of foreigners" who supposedly attempted to rush boats occupied by women and children.[43] The *Armidale Express and the New England General Advertiser* of New South Wales ran a story called "A Great Canadian Calamity" that described piles of dead, naked bodies and robbers who "have been going among the dead stealing clothes and belongings."[44] Others described a "train load of coffins,"[45] men "forced to [fight the flames] at the point of revolvers,"[46] huge waves pulling victims under the lake water,[47] and "trails strewn with the bodies of prospectors."[48] Not all the Australian and New Zealand papers were so sensational. The *Daily Post* sensibly quoted information from the Ontario government, particularly Minister of Lands and Forests Frank Cochrane who stated that "while he did not want to minimize a national calamity, there was every justification for the feeling that newspaper reports of the horrors at the Porcupine Gold field as a result of the forest fires were much exaggerated."[49] Most of the missing had, in fact, been accounted for, and the total deaths "do not exceed 75." Cochrane speculated that misinformation stemmed from the difficulties in getting news out of Porcupine, the fact that some casualties were accidentally counted multiple times, and that the missing were reported dead preemptively.[50]

Porcupine's importance beyond Ontario is apparent in the longevity of the story in the press. Updates continued well into the fall. The Porcupine fire appeared, for example, in the *Gundangai Times* on September 29, 1911, in New South Wales, Australia, with many of the trappings of earlier tellings.[51] The *Daily Colonist* in British Columbia used Porcupine to caution white sportsmen against smoking in the bush.[52]

Another reason for wide interest in the Porcupine fire was that people all over the world had friends and relatives there. By 1911 the area boasted a diverse population. The town of Porcupine had 5,391 residents (divided between Porcupine North and South). Most residents (67 percent) were born in Britain, Scotland, Ireland, or France. However, there were also large groups of Russians (11 percent), Italians (6 percent), Hungarians (4 percent), Germans (2 percent), and Scandinavians (2 percent), as well as scattered numbers of African American, Chinese, and Indian residents.[53]

Porcupine's demographic diversity can be corroborated from the "death lists" assembled by the Porcupine Relief Committee. The committee telegraphed these lists to international media outlets and the Canadian government for the purposes of identification. The lists include a name or nickname and (in most cases) a place of origin for each individual. Some also included employer and next-of-kin information, along with random details of people's lives. Although the lack of consistent format precludes statistical analysis, a few rough patterns emerge. As in the census, people in the death lists came overwhelmingly from the British Isles, followed by the United States and western Europe. However, both the census and the death lists also include relatively large numbers of eastern Europeans from Poland, Finland, and Russia—the beginnings of a labor force that would grow significantly over the next two decades.[54] In 1911 risk remained relatively evenly dispersed among these groups. During the 1911 fire, managers were no safer than laborers and Englishmen no safer than Poles (of course, wealthy investors, connected to Porcupine only by their pocketbooks, were safest of all).

Aside from the census and death lists, a collection of letters adds texture to Porcupine's connectivity to the world at the moment of the fire. When people came to Porcupine, they brought the accruements of their previous lives, including familial and relational ties. Letters to and from Porcupine speak to these relations. One undated letter from around the

TABLE 1. Porcupine Population (1911)

Origin of the People by Subdistricts	Porcupine North and South
British English	693
British Irish	1,029
British Scotch	640
British Other	10
French	1,214
German	100
Austro-Hungarian	237
Belgian	6
Bulgarian and Rumanian	57
Chinese	20
Greek	5
Indian	43
Italian	340
Jewish	67
Negro	3
Polish	78
Russian	599
Scandinavian	107
Swiss	4
Unspecified	139

Source: Canada Census, 1911

time of the fire reads, "Well mother Dear I am always on the lok [*sic*] out for Lils nice letters. . . . [W]ith best love to you all and trusting you are all well, From your loving son."[55] Another refers to a missed dinner last time the author had been in the city.[56] One, on the back of a post-card showing a wooden shack, reads, "Dear Mother, how would you like to live in a house like this?"[57] Communications surely flowed regularly between miners in Ontario and their friends and families without formal documentation—but the fire ensured that for a brief, terrible moment in time, these connections would be enshrined in the documentary record, granting us a selective snapshot of the lives of Porcupine residents and their ties to other places.

This is because, in the absence of communication from Canada, and often having read alarming newspaper articles about the fire disseminated by their local papers, relatives of Porcupine residents wrote to the government for news. Secretary of State Thomas Mulvey and the Canadian high commissioner to the United Kingdom were inundated with letters from worried residents in the weeks after the fire. From July until September, when the letters finally stopped, Mulvey acted as middle man between the Porcupine relief committee on the ground in the North and family members in Britain, Australia, the United States, and Europe. The letters chronicle the emotional networks tying Porcupine to the world in 1911. Among these authors, Mary Coutts of Stovehaven, England, wanted Mulvey to "please if possible let me know the fate of my Sister, Mrs. F. Wilde. . . . [S]he opened a restaurant or hotel in [Porcupine] this spring. . . . I have one of her boys at school."[58] Mrs. P. T. Bolan of Newcastle Upon Tyne wrote to the secretary on July 17 to ask after her husband, who had been working on the Asquith Claim. "I am nearly distracted to read of the terrible fires in South Porcupine," she wrote, "and fear for my husband."[59]

In addition to sending general telegram updates on Porcupine's status after the fire, the relief committee began notifying relatives of the deceased. Many relatives heard nothing or disbelieved the reports.[60] Mrs. K. V. Taylor, after receiving a telegram at her home in Berkshire, England, notifying her of her husband's death, wrote to Mulvey to ask if there had been a mistake and that she "should like further evidence that this William Taylor was my husband." She provided other details about his life that might be used to identify the body. The stakes for Taylor were high. She had sold her home to pay for her husband's passage to the mines and had planned to join him in Canada when he found a job. "If you could inform me in any way I should be thankful as myself and three children are left destitute," she wrote. When her husband was confirmed dead, Taylor requested her name be given to the Relief Committee for possible financial support.[61]

Those who did not receive official notices read familiar names on the death lists and worried. For example, the *Daily Telegraph* listed "M. Martin, Thoringey, Los Angeles" among the dead. One Mr. Stanley Moon wrote to Mulvey with a copy of the clipping. His family friend Miss Thoringey arrived in Quebec on July 10 from Liverpool, and Moon, unfamiliar with Canadian geography, worried she'd ended up in Porcupine. She had not—as Mulvey testily wrote back. "Mr. Martin Thoringrey" was in fact a consulting engineer from California, and "in any case it is extremely improbable that a woman landing at Quebec on the 10th of July would have gone into the Porcupine district."[62] In another example Ann Hogben of Folkestone, England, wrote to ask for clarification regarding her son Arthur Dexter. The name "Charles Dexter of Folkestone" had appeared in the death list, and Hogben "feared that the name given, Charles, was merely a mistake made in the general confusion."[63] If her son had indeed died, Hogben wanted to know what had been done with his body and belongings and whether she could recover them. Her request to Mulvey was forwarded to the Relief Committee, which sent additional details and confirmed her son's death.

Porcupine's destruction would have widespread impact on a collection of geographically distant lives. Mining districts are often associated with populations of solitary men, disconnected from family, and operating in homosocial environments. This perception obscures complex gendered relationships on the ground in emerging resource communities: historians have shown how working and racialized women and men formed complex internal social structures and hierarchies that shaped work and life.[64] The letters from the Porcupine fire reveal Porcupine's personal, emotional, and familial ties to a community of anxious stakeholders around the world. In addition to the internal connections they built on the land, miners at Porcupine remained connected to an external relational world that exerted influence over their lives even at a great distance. Men and women at Porcupine lived and worked within gendered, familial, and social structures both within and outside the mining camp.

Outgoing telegrams from the relief committee contain hints about the experience of Porcupine citizens in the days immediately following the fire. Survivors cared for the injured and combed the debris for bodies. Many were severely burned (one man had his feet burned off), and some died in transit to hospital.[65] A telegraph from Porcupine dated July 13 states that "tents blankets and provisions urgently wanted. Conditions underestimated." Periodic reports suggest an ongoing search for bodies, many of which were buried in place by survivors. One telegraph reports, "Mike Johnson found on Trout Creek Night Hawk district committed suicide," and "remains of man and Spaniel found under Theatre at South Porcupine supposed to be 'Cripple Creek' an old Western prospector."[66]

The trauma of the fire is partially captured by the Henry C. Peters photographs. A local photographer based in Porcupine, Peters documented the fire and its aftermath thoroughly. In Peters's photographs, men are pictured carrying bodies on stretchers on the pier at Porcupine Lake and standing next to rude graves and stacks of coffins. There are

At least seventy-three people died during the 1911 fire. Some victims were buried in place while others were interred at Dead Man's Point, a rocky outcrop on the south west shore of Porcupine Lake. Toronto Star, "Graves of Victims of Great Forest Fire 1911," Box 04564, MIKAN no. 3298974, Library and Archives Canada, Ottawa, Ontario.

Photographer Henry C. Peters documented the destruction caused by the 1911 fire, including this image of the mine shaft where dozens of sheltering miners and their families suffocated. "Mouth of the West Dome Shaft in which Captain Weiss and Family with 27 Companions died," CPC-00707, the William Ready Division of Archives and Research Collections, McMaster University Library.

images of people wandering the ruins of the town and one of two men sitting next to the burned-out Dome Mine shaft where Weiss and his family died. Many were made into photo postcards as a way of memorializing the fire and communicating its destructiveness to distant audiences.

Negotiating Nature and Labor in the Return to Mining

As traumatic as the fire was, it took little time for the practical work of mining to pick up where it had left off on July 9. Indeed, in the long run, the fire did little to alter the trajectory of extraction toward large-scale, capital-intensive forms, or the "rich man's venture." It may have even expedited the process by clearing land and by financially destroying small operations with few outside resources and tenuous holds on title. Yet the reconstruction of big, global industry on scales larger and better connected than ever before still failed to immunize Porcupine against environmental disaster. At the same time, the fire undid a considerable amount of human work on the land and delayed development that otherwise might have occurred more quickly. Following the fire many of the survivors returned indefinitely to homes and families elsewhere.[67] On a basic level, mining companies had to rebuild the infrastructure they had lost before they could consider expansion. In several cases, companies had been waiting for completion of the railway to bring in major equipment.[68] Hollinger had been waiting for the railway to bring in tube mills and cyanide.[69] The fire added to existing delays in rail construction, and companies waited longer than anticipated for materials. There were some losses for the big companies, and managers struggled to make the most of a bad situation.

There was also the somewhat awkward matter of the in-place burials near Porcupine Lake. Many bodies had been interred on Dome property, to the consternation of management. The site was on a high piece of rock overlooking the lake and was the exact location where the first prospecting crews had camped in 1910. It was potentially valuable land for mine infrastructure. By the end of August, Dome's management was forced to accept the loss of the land occupied by the little graveyard. In a letter to mine owner William Edwards, mine employee G. S. Harkness explained that there were too many bodies for easy reinterment. Besides, locals already called the place "Dead Man's Point." The gruesome association, he proposed, made it unsuitable for construction, especially for a residence area or a hospital. However, he assured his employer that rock in the swampy area was unlikely to contain gold. The association with the fire combined with natural characteristics to render the land useless to Dome. Harkness eventually negotiated with the Relief Committee to survey the lot for sale to the Porcupine community.[70]

For the big companies, time was money. Salvaging what they could, they went back to work at the first possible opportunity. The fact that the railway had not yet been completed became a blessing in disguise, since it meant many supplies and pieces of equipment were sitting safe in southern depots on the day of the fire. By August optimism (or at least boosterism) returned full force to the Porcupine camp. An article from the *Porcupine Press* just a month after the fire made a list of big names in the mining industry currently at Porcupine and used their presence as evidence that "investors can be assured of the importance of the Porcupine camp because of the importance of the men now there." Names included Captain DeLamar, who had worked in Colorado, Idaho, Nevada, California, and more recently in Cobalt. Also at Porcupine was the Bewick-Moreing Company, based in London but best known for its contributions to Australian mining, and John Millikan, owner of Golden Cycle Mill in Gold City, Colorado, famous for treating a thousand tons of Cripple Creek ore per day.[71]

Although the fire represented a major financial and emotional blow to the mines and communities at Porcupine, there were some benefits to the burn-over. Specifically, the blaze cleared obscuring ground cover, exposing shield rock. Geologists and mapmakers got their first good look at the underlying formations. Harkness wrote of the "point" claims (those around Dead Man's Point) that "the fire has so swept these claims that you have a much better chance, now, to pick out the good spots, than formerly."[72] Similarly, on the Whitney claims, he wrote that "the Claims are burned clear of moss, so that any outcrop of rock can be plainly seen."[73] Pesky overburden, which stymied the work of geologists, prospectors, and surveyors, suddenly was not a problem in the late summer and fall of 1911. Yet this advantage was not experienced by everyone—Harkness and the Edwards syndicate were in a unique position. Smaller companies did not have men on the ground and could not return to exploration before the snow returned to cover the rock again. Financially devastated and unable to capitalize on the small opportunities afforded by the fire, these companies began to fall behind the bigger organizations until they were eventually bought out or forfeited their claims, often after years of stagnation and financial trouble.[74]

Nor did fire-exposed rock necessarily guarantee easy profits for companies. Gold still failed to appear where it should have, veins narrowed or broke off or changed direction unexpectedly, and snow threatened to curtail prospecting and cover visible evidence of any claim's richness (essential for securing investors).[75] In 1911 Dome employee W. G. Allan noted that he was "pulling out the quartz which makes a nice showing on the dump," but that he did not want to assay as "I am getting no values." In other words, when showing the property to potential buyers, he wanted to have big, visible piles of quartz (popularly known for its association with gold concentrates) to make the claim look good—even though Porcupine quartz often contained no gold. At the end of December, Allan wrote that the face (where he had been taking out ore) seemed to have run out of gold, possibly because "the vein has shifted over" due to faulting. He was reluctant to sink the shaft farther (at additional expense) until he saw some sign of gold. Furthermore, he complained that "our progress is very slow" because "the north crosscut is the Keewaitin and very hard rock."[76] Profits from the fickle rock were not assured, and optimism whether well founded or not was part of a larger effort to build the enterprise.

In the postfire era, the big mines turned to international expertise to understand and exploit its local rock.[77] Dome hired William W. Mein as consulting engineer in 1911. Mein "had a large experience in the development of gold mines, not only in America but also in South Africa." They also hired Henry Hanson as mill superintendent. Hanson came from San Francisco, where he'd worked for Merrill Metallurgical Company.[78] The expertise of these men shaped their decisions when setting up and operating the new equipment.[79] In light of better understandings of Porcupine's low-grade ore and improvements in extractive technology elsewhere, especially from the United States and South Africa, companies rebuilt mines on ever-larger scales. In the 1912 Hollinger annual report, Edwards told shareholders that a forty-stamp mill with tube mill and cyanide equipment was well under construction and estimated the new equipment could handle 350 to 400 tons daily. They had made room for expansion, and the new powerhouse could operate a mill with double that capacity if needed.[80] The idea was to maximize the amount of ore that the plant could handle, because greater tonnages would reduce marginal costs.[81]

Porcupine mines looked for international experts and technology, but they also looked at international experience with ore comparable to that found locally. In the Hollinger annual reports, mine manager P. A.

Robbins argued that similarities between Ontario's geology and international geology and cooperation between mining fields could bring mutual benefit. "We are located in one of the great pre-Cambrian areas of the earth's crust," he wrote, so "the results of mining in similar rock formations cannot be ignored" when it came to developing Hollinger's strategies. He referred to a report presented at the Canadian meeting of the International Geological Congress that year by Dr. Malcom Maclaren, who had discussed the buildup on knowledge on similar rock from Western Australia, southern India, South Africa, and Brazil. The most important finding, in Robbins's opinion, was that gold deposits of the type located in Porcupine and these other places tended to be "deep seated in origin and persist in depth until some unfavorable change in rock formation occurs to adversely affect the gold bearing lode." Robbins was quick to point out that, at present, they were a long way from any such geological change and that diamond drilling had shown no change to the deposit would occur in the foreseeable future.[82] The deep-seated nature of the deposits was what was important. Foreign geology reinforced the notion that Porcupine mining was not unique, the lesson being that profit would come from going down rather than out. Mining companies looked to international engineering and geological science for guidance on how to manage their local deposits, and in doing so tapped into established scientific networks spanning the world's mining continents.[83]

Porcupine's postfire reconstruction was dominated by large companies whose international connections helped them to endure the disaster, take advantage of the small benefits it afforded, and rebuild quickly. Based on engineering and geological knowledge of international experts, postfire rebuilding prioritized large-scale, low-grade operations. In 1909 prospectors and investors had spent a lot of time improvising on a landscape with an uncertain future. After 1911 they forged ahead with a much clearer vision of how to make the ores found at Porcupine profitable and a rich body of experience from around the world to help them get there.

Although they recovered relatively quickly, the big mines came out of the Great Fire of 1911 with a new respect for the power of nature. In his annual report for Dome in 1912, President William Edwards assured his readers that all the rebuilt mine buildings were "absolutely fireproof." The mill, rock house, powerhouse, storehouse, laboratory, and even the houses, hospital, clubhouse, and kitchens were made of concrete or brick and steel.[84] He also noted that, when he chose the site for the new mill,

Flooding followed the Great Fire in the fall of 1911, washing away rail lines and houses. "Wash Out on T.N.O., South Porcupine, Ont.," CPC-01797, the William Ready Division of Archives and Research Collections, McMaster University Library.

"it was evident that protection against bush fires was necessary." Thus, he chose a spot protected by Miller Lake to the west, Gillies Lake to the north, Pearl Lake by the east, and to the south "one half mile of cleared ground. We propose to cultivate the ground in the spring as further protection against fire."[85] The recent destruction made managers and mining investors nervous, and they worried openly about preventing future catastrophes in the face of rising insurance costs and apathetic forest-protection measures.

Fireproofed infrastructure did little to help the mines adapt to the subsequent flooding problems. Deforestation (from both the fire and the mining activity) likely contributed to low absorption, soil instability, and increased water flow over the landscape. Clearing "overburden" for early exploration and mining meant that a lot of good timber in Porcupine had already been cut by the time the Great Fire obliterated whatever was left in the immediate vicinity. At the end of August, northern Ontario entered a rainy fall, resulting in floods and mud. Dome employees on the ground described the difficulties caused by the excess water. Harkness wrote that "the weather has been very backwards of late; it has rained nineteen days out of the last twenty-one days." On October 8, 1911, Allan wrote that the shaft "makes a great amount of water." He was required to "keep two men on a shift to keep it clear"—a major expense to the company. One week later, he commented again that "we are developing considerable water which taxes our little plant."[86] Water in shafts plagued other companies, too. In September 1912, the *Porcupine Advance* reported that "the Shumacher mine at Pearl Lake has been flooded and for the present put out of business." The accident had occurred when miners blasted too close to the edge of the irregular layer of sand surrounding Pearl Lake, allowing lake water to flood in.[87] Photographs show men attempting to repair a flooded railway, floating houses near the lakeshore, and the newly completed Porcupine express on rails totally submerged by water.[88] Fall was followed by an early winter: By October 8, there had been enough of a cold snap that Allan began winterizing the mine.[89] By the end of the month, it had snowed so much that he returned to Haileybury to get his winter gear, "as everything here looks as though it had set in for keeps."[90] Lack of timber likely contributed to flooding. The mining companies were forced to go far afield for their timber by 1912. That year Hollinger towed fifty thousand feet of mine timber up the Mattagami River, cut mostly from Bristol township and "territory north of the Grassy River."[91]

In the aftermath of the fire and subsequent rise of large-scale industrial extraction, Porcupine was forced to confront questions about the nature of mine work. The death of so many people during the fire raised questions about worker and community safety. Underground work was physically and psychologically difficult—miners called their conditions "unnatural and trying," as they were "away from the surface and the sunlight and in air more or less impure and inferior to the natural air." Furthermore, air in the mines was "often contaminated with fumes of gas with injurious dust of particles from drilling and other operations."[92] As work became more monotonous, difficult, and specialized, Porcupine workers worried about the health of the mine environment for their bodies. Porcupine miners joined the Western Federation of Miners starting as early as 1910, but the rising tensions between workers and companies indicate a change in the nature of work at Porcupine after 1911.

The eight-hour day was among these new tensions. Conflict over working hours had led to strikes elsewhere in North America and resulted in legislation in British Columbia as early as 1899. In 1912 Samuel Price investigated the eight-hour day for the Ontario government and found miners nearly universally in support.[93] Porcupine miners paid attention to international conversations about working hours, and conversations about an eight-hour day started in the months following the Great Fire. Many pointed "to the various other mining countries where the 8-hour law is in force" and asked why Ontario "should not be as good as these."

They argued that skilled miners from the western United States and British Columbia might be attracted to Ontario with an eight-hour day and dismissed the "cries of injury to the industry" by pointing out that such concerns internationally "have proved ill-founded."[94] Efficient management would mitigate any potential losses in time (the same amount of work now completed in ten hours could be completed in eight, miners claimed, with better planning).[95]

Mine managers had also been paying attention to the rise of the eight-hour day in other districts and recruited evidence from abroad to argue against its implementation. Price quoted managers who said that the eight-hour day "has in fact had disastrous effects in British Columbia, Australia, and parts of the United States, and is driving capital and labor out of England and the United States." His sources further blamed agitation from "undesirable foreign organisations" working for their own advantage. The fact that the Ontario mines competed on the global market was also used as a reason against implementing the eight-hour day.[96]

Price checked the evidence, did the math, and found that, contrary to the arguments of Porcupine's mine managers, the eight-hour day did not actually decrease output in other states.[97] He recommended in favor of the eight-hour day in his report, and the government passed legislation to enact it the following year.

Porcupine's cadre of newly powerful industrial mining companies was unhappy with this conclusion, which they saw as a direct threat to the economies of scale required for profitable extraction. They accused Price of being an advocate rather than a judge and asserted that men were generally happy with longer working hours. An editorial in the *Canadian Mining Journal* asked why a "paternalist" government would interfere with mining (in which everything appeared to be going quite well) when there were so many other industries where workers were treated more unfairly. Contrary to the workers' concerns quoted by Price, the companies argued that underground work was "no more 'unnatural' than work in any building, or on a ship, or on a locomotive. . .and. . .it is infinitely more wholesome and much more remunerative than the majority of vocations." As for concerns about air quality, "we may frankly say that we do not believe it." The journal also argued that Price had overlooked the substantial *benefits* of underground work environments: "The miner is not exposed during working hours to inclement weather and to variations of temperature."[98] Especially in Ontario, the stability of temperature deep underground might be seen as desirable when it got cold in the winter.

Hours were not the only place where miners chafed against the new requirements of industrial labor. The end of 1912 also saw violent strikes following wage cuts in the fall of 1912. The mining companies eventually dealt with the first strikes in 1912 by bringing in labor from outside the camps—largely inexperienced and unskilled men who took time to train.[99] When the strikers tried to block access to the mines, Hollinger hired private constables to disperse them. The strike was called off in 1913 but was cited as a reason for low production for both 1912 and 1913.[100] Letters from May 1913 indicate that labor remained scarce that spring. One of Edwards's managers wrote that "it would not be advisable to commence work on the Edwards Lot at present owing to the complicated labor situation. No good miners in camp."[101] The comment reflects the region's inability to attract skilled workers at prevailing wages and conditions, a common problem among Porcupine's contemporaries in South Africa and the United States.

Were workers' concerns about large-scale industrial mining environments justified in the years after the fire? The work in the post-1911 mines could be dangerous. The 1913 annual report for Hollinger Mines records six cases of grippe, four of typhoid, four lung and throat disorders, three nerve disorders, two intestinal disorders, and three cases of lead poisoning. The report also listed nine men hit by falling rock, six injured while handling mine cars, ten hurt by falling tools, three hit by dropped cages, two who stepped on nails, and six who fell from scaffolds. Men also slipped, were burned by boiling grease, hit by trees, fell from wagons and moving cages, and were injured while working with the compressor, shifting belt, acetylene gas, splinter, hoses, and spikes.[102] The accident report listed three fatalities: One man got caught up in mill machinery, which crushed him to death. The other two were caught in an accidental dynamite blast.

The annual reports for the Ontario Bureau of Mines suggest that little had changed to improve mine safety as the industry scaled up. In 1912 Deputy Minister of Mines Thomas Gibson wrote that accident statistics "reveal little or no improvement over previous years. The casualties which kill or maim so many of the men engaged in mining in Ontario are too numerous." By 1914 things had not improved, but Gibson blamed the poststrike hires, many of whom were eastern European, lacked mining experience, and often did not understand English.[103] The 1914 report also discussed miners' phthisis for the first time—a lung disease caused by the inhalation of silica dust by then endemic at the big industrial mines of Western Australia, South Africa, and the United States. Over the next decade, phthisis, or silicosis, became a health crisis in Ontario. The mines that rose from the ashes of the 1911 fire contained a host of new hazards for workers that would become only more extreme after the First World War.

The health impacts of a burgeoning industry were not restricted to mine premises. In August the *Porcupine Advance* carried a story about "Indians living up the Montreal River, in the direction of Fort Matachewan," suffering a diphtheria epidemic. The paper revealed that the group normally traded at Elk Lake post, but they had been barred from entering Elk Lake in an attempt to limit the spread of disease.[104] The epidemic came in the context of increasing incursions on Indigenous title between 1911 and 1914. In 1911 a syndicate represented by Cyril T. Young wrote to the Department of Indian Affairs to request access to the Matachewan Indian Reserve. He contended that, under the control of the "Indians," the land would never be explored or developed, and besides, he argued, Treaty Nine was unclear on who controlled minerals on reserves. He wrapped up his letter with an offer to pay unspecified amounts of money for access.[105] Indian Affairs promptly rejected the request, stating that the newness of the reserve made prospecting in the area particularly undesirable and that the department would not allow any white person to occupy or use the reserve lands.[106] Young persisted, indicating that he could easily get the consent of the "Indians" if that was what was needed, but Indian Affairs again rebuffed him, stating that the "morality" of the "Indians" needed to be considered and that there were other (unspecified) reasons that prospecting would not be allowed on Treaty Nine reserves.[107]

Consolidation and Geological Adaptation, 1913–1915

Despite accumulating environmental and health concerns in and around the growing mines, the trend toward corporate consolidation under the "big three" accelerated in the years leading up to the First World War. By the time that the 1913 annual report of the Bureau of Mines was released, hardly a trace of the fire's economic scars could be seen on the big mines' record books. In 1911 Ontario had produced $42,637 worth of gold. In 1912 it produced $2,144,086, most of which came from the Porcupine—a huge jump that did not even represent a full year's work.[108] In 1913 totals more than quadrupled to $9,293,231—not for all of Porcupine, but for Hollinger and Dome specifically.[109]

At the same time, the number of licenses sold was decreasing, and it was becoming apparent that small-scale independent prospecting was over for Porcupine.[110] The high-grade surface deposits that might alert a single prospector to the presence of paying ore were largely gone. Exploration now meant large-scale drilling and coring operations employing teams of professional engineers and geologists. The numbers and profitability of smaller mining companies continued to plummet leading up to the First World War. In the 1912 annual report, Foley O'Brien was described as carrying on work until July 1911, "when all work ceased."[111] Pearl Lake and Jupiter Mines both stopped or slowed production in these years too and were eventually bought out by McIntyre Mines.[112] That year the *Canadian Mining Journal* ran an article titled "casualties" about Pearl Lake and the Rea mine. "Two deaths, one by asphyxiation, the other by inanition, have occurred in the Porcupine. The Pearl Lake mine has been shut down, smothered by over-ground exposure. The Rea mine (we believe that this is, after all, a case of suspended animation), the sound of the drill is no longer heard."[113] The journal blamed poor management, by which it meant limited or small-scale development rather than large-scale, capital-intensive, and internationally minded mining.

Whereas in 1911 the bureau counted and described thirty-four operating companies, by 1914 there were only nineteen, many of which were owned by one of the big syndicates. The annual reports do not give a clear reason for this change, stating simply that "a number of smaller companies were forced to suspend operations, and, as a result, development work has been considerably retarded."[114] A combination of the fire, labor troubles, and the low-grade capital-intensive nature of the shield deposits forced these smaller companies out of the business. Only those who had set themselves up for success by controlling large amounts of land from the beginning, and who had the capital to follow through with the necessary adaptations to the particular form of low-grade mining required on the shield, would survive.

The diversity of company names listed in the annual reports had, from the beginning, obscured the fact that many individuals had interests in multiple companies. This had not been done with the thought of large-scale, low-grade mining specifically in mind, but was a strategy for mitigating risk—if one claim did not produce paying values, the syndicate could fall back on its other investments. However, it turned out that the nature of Porcupine gold made the scattered landscape of multiple companies less efficient than large, consolidated operations. Because the gold existed in low concentrations in the rock, companies needed to pass large amounts of rock through the mills in order to make a profit. By centralizing milling operations, companies could cut down on equipment, construction, transportation, and labor costs. Furthermore, many of the veins extended across multiple properties, creating costly and inconvenient bureaucratic barriers. Once these facts became clear, and once it became obvious that the ore bodies extended deep enough to sustain the kind of capital investment needed for low-grade extraction, companies consolidated. In 1915 the bureau argued that prospectors should not bother looking for high grade veins, which "in this area have not been worked with success," but should concentrate their attention on finding the more profitable low-grade deposits.[115] The industry's commitment to working with low-grade ore in the years leading up to the First World War was an adaptation to shield geology and environment. Smaller companies and independent prospectors were the casualties.

By zooming in on the Hollinger Mine, we can see how consolidation functioned at a practical level. From the beginning, the syndicate backing Hollinger had its fingers in a number of pots. Not only did it own the four claims originally staked under its license, but it had also obtained seven adjacent claims. These they had broken down into three separate holdings: Acme Gold Mines, Millerton Gold Mines, and an undeveloped claim, #13147. The plan had been to develop them modestly and separately, while the syndicate spent most of its resources developing the more promising Hollinger property. Thus, Hollinger was the only property that had built (and then rebuilt) a mill. In 1912 it treated 138,148 tons of its own ore in addition to treating 1,840 tons for its neighbor Acme Gold Mines, at a charge of three dollars per ton. This strategy ensured that properties like Acme could develop slowly without the financial risk involved with putting in infrastructure on a property that might fail to turn up much gold; meanwhile, Hollinger benefited from a small extra income for processing its neighbors' ore.

In 1914 the syndicate changed tack. "Joint operation," the annual report argued, "would eventually prove the most satisfactory and economical way of dealing with the properties."[116] Appended to the 1914 report was an additional information booklet recommending the consolidation. This booklet gave two main reasons: first, "in order to derive the best of lowered working costs," and, second, "to mutually insure each against the vagaries of geological phenomenon."[117] Both justifications boil down to mitigating geological risk. The deposits were unpredictable, large, low-grade, and covered in a considerable amount of "overburden" or drift. Yet if some claims contained little valuable ore, they could still be profitably mined using the mills and refineries at proven properties. If all else failed, they could be used for other purposes that would support the profitable properties, such as providing space for outbuildings or waste storage.

The report began with an explanation of the local geology. "This old series of rocks has been folded and crumpled and torn until its original stratigraphical features have been effaced, and chemical changes over the course of time have so modified the rock forming minerals that there are now left practically none of the original constituents." The resulting landscape was one of high ridges, where the glacially smooth rock was exposed, and low valleys covered in drift. This was less than ideal. Yet management had "found with development" that the remaining deposits "are more or less connected together in chains, forming ore bodies hundreds of feet in length."[118] Thus, the Porcupine deposits could be rendered only on a large scale.

The report went on to state that the legal requirements for claim staking were not particularly well suited to the geology. In short, management found that nature did not conform to established political boundaries: "While the legal requirements for the staking of claims have been left to the fixing of arbitrary boundaries, there are no natural boundaries between the properties." In fact, the "arbitrary boundaries" impeded efficient exploitation because they tended to cut the ore bodies in inconvenient ways, which made amalgamation the more sensible, perhaps even a more "natural," path.[119]

Although Hollinger's management understood that the ore bodies were large and long, they did not know much else about them. Rock was fickle, and although it obligingly provided tons of gold-laden rock for now, its generosity could very feasibly end at any moment. Any faulting or other irregularities in the rock could cut off an otherwise profitable ore body. The idea made mine engineers and investors nervous. If this should occur, it would "entail a severe setback to one property and a consequent gain for another." Consolidation would solve this dilemma by taking the uncertainty out of geology. Besides, a more complete understanding might be gained by looking at geology as a whole, instead of artificially divided parts.[120]

The argument for consolidation laid out, the report turned to a description of the practical work involved in arranging it. The Acme claim had already been developed to some extent (because it contained some high-grade ore), but Millerton and claim #13147 had been barely touched. The two were covered with a considerable amount of drift. South Millerton was covered with "a heavy overburden of sand which has prevented prospecting," and the north had shown only low grades. Part of the property was also covered by Miller Lake. However, if the Hollinger veins continued without interruption, the report speculated that they would eventually extend onto this property. Claim #13147 remained totally untouched, but Vipond mines, which was adjacent to the property, was making profits, and it seemed likely that its deposits might extend onto the property.[121] So although there was some risk inherent in taking on the unproven and mysterious Millerton and claim #13147, there was some indication that the gamble would pay off for investors. Besides, the proven Hollinger and Acme claims could profitably exploit low-grade finds, whereas an independent Millerton and claim #13147 could not. Consolidation would be driven by the desire to mitigate geological risk. The characteristics of the rock meant that small companies were exposed to more risk and higher operating costs than larger companies. By amalgamating, Hollinger's management hoped to reduce the risk of their veins ending unpredictably while lowering the costs by putting more low-grade ore through central large-scale machinery. As the Hollinger example shows, the takeover by the "big three" represents an adaptation to the conditions of the Porcupine deposits.

Fire Fortunes, Fire Failures

Colonial practices of fire and brush management in northern Ontario were unsustainable, and the fire of 1911 occurred at their breaking point. Insect kill, windfall, and warm climate combined to make the North particularly susceptible to a large fire. Fire suppression, a lack of fireproof infrastructure, easily broken transportation and communication networks, and an inability to anticipate the fire worsened the situation and resulted in surprise, death, injury, and confusion on the day of the fire and in the days and months following. The Great Fire disrupted the human and environmental history of norther Ontario, but it also changed the trajectory of uncountable lives connected to Porcupine by a variety of threads—monetary, emotional, and familial. The fire cleared the landscape of most of the visible signs of Porcupine's connection to an international mining community and flattened most signs of progress erected or excavated since 1909. Reminded of the power of Canadian Shield nature, companies changed their building practices and went to work assembling more resilient infrastructures—none of which helped with the floods and labor unrest that would disrupt extraction in the months afterward.

Yet in the years that followed, the fundamental trajectory begun in 1909 remained unchanged. If anything, the fire functioned as an accelerant to processes well under way, emphasizing the extent to which it was fundamentally the product of gold mining rather than external to it. Before 1911 the uncertain future of Porcupine had resulted in slow and sporadic development and a mix of small surface and large-scale mining. Afterward, it became clear to the companies and their investors that profit would come from large-scale industrial extraction on par with similar projects in Australia, the United States, and South Africa. The fire physically and ideologically cleared the landscape in preparation for Porcupine's transformation into a booming industrial hinterland by clearing the land of overburden, weakening smaller syndicates, and providing opportunities for consolidation. In a messy new ecology that grafted mining onto existing environmental systems, the Great Fire signified dysfunction. Its destructiveness was exacerbated by the fragility of the physical and social infrastructures people had built at Porcupine since 1909. Rather than destroy those infrastructures, the Great Fire honed and hardened them. Porcupine's strongest core had always been its foreign-capital-backed corporations, and it would be these ventures that recovered fastest.

Over time the legacy of the Great Fire for most people was its role as a baseline and a measuring stick for Porcupine's industrial progress. Ten years after the fire in 1921, the Great Fire was invoked in the *Globe* in a story about new fires in the North.[122] In 1935 Gibson (then minister of mines) wrote an article titled "Drama and Romance Form Background of Mining Progress: Foundations of Great Industry Laid amid Tales of Accidental Discoveries Followed by Struggles of Visionary Pioneers." Rather than seeing the disaster as a convergence between the natural, built, and social environments, these accounts pitted the fire against human progress. By midcentury, *Maclean's* magazine saw the mines as a literal manifestation of industrial resilience: "The men of the Porcupine erected no monuments," a 1954 article reads, "and then they were too busy building mine shafts, whose tin roofs pointed to the sky and whose insides dropped into the earth for gold. And by then they had built the Porcupine, and the need for monuments was past."[123] The events of 1911 became a symbol of the danger, struggle, and precariousness of human ambition on a wild frontier: people versus nature. Porcupine residents' ability to rebuild in the face of adversity, thrive in the wilderness, and realize their industrial destiny became part of the mining community's identity. Its industrial accomplishments shone even brighter when contrasted against the dark chaos of the Great Fire.

In the twenty-first century, the symbolic and rhetorical power of the Great Fire remains undiminished. In 2011 a ceremony was held at Dead Man's Point (where many of the victims were buried) in honor of the fire's one-hundredth anniversary. In a local news story covering the event, the fire was described as "a local historical milestone—it helped shaped this community and stands as an example of the wild west, pioneer spirit that created the Porcupine Goldfields."[124] In 2012, under threat from the "Timmins No. 9," wildfire locals took comfort in the fact that they had survived such calamities before and could do so again.[125]

Yet the communities and the mines continued to struggle with environmental issues after 1911, many of which were exacerbated by the increasing scale of extraction. The Great Fire was the first major crisis in Porcupine, and it foreshadowed the region's complicated future. In particular, the mines' trouble with northern Canada's hydrology combined with growing concerns about human health would escalate in the context of the power and labor shortages of the First World War.

CHAPTER THREE

No Energy for Industry

Powering the Porcupine into the 1920s

In April 1921, two industrial titans clashed in Ontario's Supreme Court. The Hollinger Consolidated Gold Mines was suing Northern Canada Power Company. Hollinger's leadership claimed they had been wronged by the rain—or, rather, its absence. Specifically, lack of precipitation and unexpectedly small snowpack for consecutive years had resulted in a noticeable decrease in the province's hydroelectric reserves. With too many contracts to fill, low waters on the Mattagami River, and an impending power shortage, the Northern Canada Power Company consulted with its mining customers on a scheme of equitable distribution of its remaining supply based on past usage. They framed it as a regrettable but necessary step in the face of disappointing environmental realities. In Hollinger's eyes, however, these restrictions evidenced Northern Canada Power's negligence. In offloading the economic costs of low rainfall to the mining companies, Northern Canada Power had failed to manage the landscape in a way that aligned with expectations for the North.

In making its calculations, the power company's engineers had failed to account for history. Since the early days of the rush, development had proceeded at breakneck speed. Having rebuilt infrastructure after the Great Fire of 1911, weathered a major strike in 1912, and started consolidation in 1914, Porcupine's mines had just begun to think seriously about how they might make their low-grade deposits profitable when the war began. As the higher-grade surface deposits disappeared, mining companies sought to extend the momentum of the rush by exploiting deeper deposits. At the time, however, neither hard-rock mining technology nor geological science had had a chance to catch up. Companies relied on out-of-date geological reports and imported tools not yet fully adapted to environmental requirements of the Canadian Shield. Despite impressive advances since 1909, the camp remained a bit of a backwater. Then the war began. Despite initial hopes for increased demand for gold, wartime money and labor shortages slowed Porcupine production.[1] The unexpected side effect of this slowdown was that the war provided time for the fledgling mining industry to organize. Specifically, wartime conditions facilitated more corporate takeovers and produced better geological understandings of major deposits. The war also materially helped the mines by glorifying metal production as a form of patriotism and, as it ended, creating a new pool of potential labor. As a result, the largest mining companies emerged from the conflict more successful, confident, supported, and informed than ever.[2] After years of hardship, conditions finally turned in their favor. The mines were ready to seize their destiny.

In the absence of the power problem, the story should have been one of unqualified success. The return of normal conditions promised the opportunity to unleash a new, more organized productive power on Porcupine at a scale not seen before in Canadian gold mines. So when Northern Canada Power made their 1921 calculations for the distribution of remaining power reserves based on *past* usage, it inadvertently put itself between powerful mining companies and their plans for unfettered industrial expansion. Past usage no longer represented mining's abilities or its ambitions. By 1920 the mining companies had been waiting impatiently to put exciting new geological knowledge, infrastructure, and financial resources to work. Hollinger in particular was on the cusp of radical expansion and had enjoyed a special relationship with the power company based on a shared history.[3] Just as Hollinger had finally shed the external restrictions inflicted by the war, it found itself unexpectedly hampered from within by those it had always assumed were on its side. Half a decade of pent-up extractive frustration found an outlet in blaming Northern Canada Power.

At the heart of the issue was the discrepancy between the mines' postwar capacities (built to compete with their large industrial counterparts in the United States and South Africa) versus the mercurial nature of shield hydrology. As Porcupine industrialized, the question of how to power the massive machinery had come up early. In 1910 the Ontario Bureau of Mines already knew the solution: "If the new gold field of Porcupine proves to be permanently workable, there will be no difficulty in harnessing the falls on the Mattagami, Grassy and other rivers within convenient reach for use in the mines." In 1910, in the context of gold-rush excitement, the proximity of the mines to ready sources of hydroelectricity confirmed the North's industrial destiny. As Bartlett commented in his 1911 report, "The importance of such water powers in close proximity to a mining camp does not call for further comment."[4] A decade later, further comment was clearly needed because the task of harnessing northern rivers to power the industrial state proved more complicated than an optimistic bureau could have imagined. Although northern Ontario possessed plenty of water, its capacity to produce power was not infinite, and it was particularly stymied by fluctuations in precipitation levels from year to year.

In 1920 the expanding mines reached the limits of what their current power infrastructure could provide. Because the North had always been framed as naturally destined for industrial expansion, the question of whether mining's power demands were even possible to fulfill was not considered. Instead, Hollinger blamed Northern Canada Power for the environment's shortcomings and called it a problem of management. With the right know-how and sufficient application of technology, power companies could overcome the apparent limits of northern nature. The court case proved that the provincial and federal governments (eventually) agreed. Aligning the landscape with industry's expectations required large-scale, expensive, and destructive modification of the North's watersheds.

Hollinger's suit ensured mining companies would not bear the environmental costs of expansion. Instead, these would be dispersed among working-class miners and Indigenous people. Cheap labor and waterpower were the twin pillars supporting industrial mining. The war had brought an influx of eastern European labor to Porcupine. These newcomers were cut off from home networks of familial and financial support and disempowered and distrusted by the Canadian state and its citizens. Mines became complex systems that blended human and hydrological energy to extract gold from the earth. Without miners' bodies to extend the labor of machines, the extra waterpower would be unnecessary. Although these new hydrohuman systems were efficient from a profit-making perspective, they had costs. While they lined the pockets of investors, massive hydropowered mines were dangerous spaces that exposed miners to industrial disease and accidents. On the land around the mines, the Northern Canada Power Company built the Mattagami reservoir, flooding a large swath of Mattagami First Nation's land; destroying valuable timber, hunting, and trapping land; and displacing people from their homes.

At first glance, the water shortages of the 1920s (and their consequences) look like a straightforward case of external forces dictating local environmental change under the twin regimes of industrialization and international capitalism to the detriment of resident people.[5] Although it is true that those who made the decisions to shift the costs of production through hiring practices and dam building did not see or experience its consequences, decision makers always relied on the knowledge, actions, and experience of people "on the ground" in Porcupine. They understood this dependence, worried about its consequences, and worked actively to counteract it. Like the Hollinger hydro suit, this anxiety was rooted in the existential threat represented by exhaustion. One day all the gold that could be profitably mined would be gone. If the mining companies wanted to push that day further into the future, they needed more power and more prospectors. Yet the two seemed mutually exclusive because hydroelectricity created mines full of wageworkers rather than the enterprising individuals empowered to explore new mineral frontiers.

Labor and Danger During the Great War

Looking back on the first year of the war in 1915, Deputy Minister of Mines Thomas Gibson wrote that "[mining] statistics. . .reflect the influence of passing events, whether on the provincial, national, or world stage."[6] This was his way of explaining why production and profits had fallen since the outbreak of war. Initial optimism about the opportunities for gold mines under a cash-, credit-, and metal-hungry war economy had given way to conservatism and caution. A significant economic depression followed by the declaration of war caused economic devastation in Canada and saw capital dry up, prices for metals drop, supply costs rise, and available labor dwindle at the mines.[7] If the war later created the conditions for improved geological understandings and mine planning, these benefits came about only because the industry was making the best of an otherwise very bad situation.

It took time for the outbreak of war to directly impact Porcupine. Even after prices for other precious metals began to fail, gold remained mostly immune because of its role in balancing international payments on current accounts.[8] Compared to skyrocketing production since the Great Fire in 1911, profits in 1914 certainly slowed, but not nearly to the same extent as silver and other metals.[9] In fact, 1914 was the year Porcupine made Ontario the biggest gold producer in Canada: Ontario mined 268,942 ounces worth $5,529,767 compared to British Columbia's $5,177,343 and the Yukon's $5,125,396.[10] The numbers were a product of the early success of the large-scale, low-grade mining model in Porcupine buoyed by the economic stability of gold.[11] Not only was Ontario's gold coming from a small geographic area, but it was produced by only eight companies.[12] In 1914 the mines were celebrating their success and did not appear worried about the prospects of the future.

This is not to say that they were completely blind to the potential consequences of the conflict. In 1914 the *Canadian Mining Journal* optimistically asserted that the war would not affect the "self-supporting" mines of Porcupine.[13] By "self-supporting," the journal referred to the mines already able to ship bullion, not those still in development (or fire recovery). In other words, larger companies with international connections and bigger bottom lines stood the best chance of success. Others would have more difficulty. Evidence of their hypothesis came early. Foley O'Brien and Schumacher, both smaller operations, floundered in the fall of 1914 as exploration and development capital dried up in the increasingly risk-averse investment climate.[14] Conversely, among the "self-supporting" mines, there was some modest expansion. Hollinger planned to increase the capacity of its new mill from 1,600 to 1,900 tons, sink a new shaft to connect its newly consolidated properties, and connect everything with electric locomotives.[15] McIntyre and Vipond mines expanded their mills.[16] Dome, meanwhile, focused on productivity gains by demanding more of a reduced labor force.[17] In 1915, Timmins congratulated his shareholders on their Hollinger investments. He reported a massive increase in the ore reserve and celebrated the fulfillment of estimates made in 1912.[18] According to the Ontario Bureau of Mines, by the middle of 1915, "the mines were worked with feverish activity in order to supply the abnormal demand for metals." For those already positioned to sell it, the price of gold went up in these years and gave the producing companies a boost at the beginning of the war. Meanwhile, the *Canadian Mining Journal* expressed anxiety about the demise of smaller operations and called for government intervention to help support the work of smaller operations.[19]

Any optimism in mining, as in other parts of the war economy, assumed that the conflict would be short-lived. As 1916 came and went with no end to the hostilities in sight, optimism became increasingly strained. Even the big mines simply could not endure the inflated prices for zinc, explosives, steel, and other supplies in the long term.[20] Cyanide (imported from Germany) became the first major worry. Low-grade gold mining requires cyanide (mixed with water) to separate gold from ore and trace metals, but in the early twentieth century only a few places manufactured the chemical on a large-enough scale for exportation. While there was some talk of an alternative supplier in Britain, the

Canadian Mining Journal speculated that most of this would be exported to the Rand, which, despite, Ontario's recent advances, still outproduced Porcupine by a considerable extent and contributed directly to Britain's financial reserves.[21] In 1916 both the Bureau of Mines and mine managers expressed concern about the steady rise in prices for mining and milling supplies.[22] "We must admit that we now have a fuller appreciation of the seriousness of the struggle in which the Empire is at present engaged," wrote Hollinger president Noah Timmins at the beginning of 1917. This new appreciation made it clear that "some readjustment of our plans has become necessary." Mine manager P. A. Robbins admitted that "we failed to make due allowance for the magnitude of the task confronting Great Britain and the Allies." Among the challenges faced in that year, Timmins listed "annoying delays in the deliveries of machinery" and the fact that "the costs of materials and supplies are still rising."[23]

As the war dragged on, the state of the labor market worsened. Porcupine's prewar model of low-grade gold extraction depended on a steady supply of cheap workers to move gold ore through the industrial machinery.[24] The mines could not afford to pay workers enough to keep them. Besides loss to war recruitment, the high wages and better working conditions available in the American West drew experienced men out of Ontario. Agriculture, given additional importance during the war, took remaining labor away in the summer and fall. Lumbering companies poached available workers in the winter.[25] Employment numbers from Hollinger give a sense of the turnover experienced at the mine. On December 31, 1917, Hollinger had 1,230 men on its payroll. It had hired, over the course of the year, 2,700—more than 100 percent turnover.[26]

The high turnover also meant that the mines were hemorrhaging men with mining experience. Mining is a unique kind of work, and it took time and practice for miners to become efficient at it. New labor worked painfully slowly, and the impact on profits was noticeable.[27] For example, the amount of drilling conducted during an eight-hour shift directly correlated with the amount of ore the mills processed. The amount of ore milled in turn directly correlated with the amount of gold. Slower drillers meant less ore, smaller outputs, and less gold. As mining and technology historian James Otto Petersen has aptly put it, using unskilled workers in the mine "became logically equivalent to operating a machine below capacity."[28] This was a major problem for an industry where profit was directly correlated with adequate volume.

The mines attempted to reduce their labor needs through mechanization, through the installation of tools like locomotives. Mechanization failed to eliminate the mines' dependence on human beings for significant portions of mine work.[29] This was true for skilled work like drilling, but it also held true for "unskilled" labor. For example, "muckers" shoveled ore from the rock faces into ore carts (belowground) and moved ore through the mills (aboveground). In 1918, Hollinger experimented with mechanical muckers, but they proved finicky, expensive, and less efficient than men with shovels.[30]

Labor shortages also put the mines in a position of vulnerability to organized action. Hollinger manager P. A. Robbins articulated this anxiety in his 1916 annual report. "It is to be hoped that [workers'] efforts to precipitate a strike will be unsuccessful," he wrote. "Under present conditions there would be nothing for us to do but curtail our operations."[31] In 1918, President Bickell of McIntyre Mines expressed the opinion that it might be better to close down the mine entirely than risk a strike.[32] In this context, the mines were constantly on the verge of inoperability.

The immediate impact of these hardships was a general contraction of human activity on the landscape at Porcupine.[33] On the recently consolidated Hollinger, Acme, Millerton, and claim #13147 group of properties, for example, mine manager P. A. Robbins stated that "shortage of labor has been a handicap to work in the mine, and has placed a limit upon the amount of development work which could be accomplished." Although the company had planned to work claim #13147, the land was largely left alone in 1916, "as we could not spare men." On the main properties, work was also affected. Mining occurred closer to the surface. Men were not sent below 425 feet often and never below 800 feet.[34] By the end of 1917, even the bigger mines began openly considering leaving the ore in the ground and shutting down the mines until favorable economic conditions returned.[35] The summary report for 1917 in the *Northern Miner* included a long list of these idle properties.[36] As it had after the early spring of 1909 and the Great Fire of 1911, the provincial government made special exceptions to mining legislation in 1917 with an Order-in-Council that allowed the postponement of assessment work on claims for twelve months. Large numbers of property owners took advantage of this break to leave their properties undeveloped.[37] The major exceptions in the shutdowns were McIntyre, Dome, and Hollinger. Rather than cease production, the big three adapted their operations by closing portions of their mines, processing higher grades, and tapping reserves.[38]

The war eventually provided a partial solution to the labor problem.

As the conflict progressed, Porcupine received wartime immigrant Scandinavians, Italians, Ukrainians, Chinese, and Balkans. Although they worked for less and enjoyed fewer rights than their Canadian and Allied counterparts, they were never able to satisfy the mines either in numbers or in skills. Chronic distrust hindered their integration into the workforce. Nevertheless, the postwar expansion of the mines rested as much on their (cheap and expendable) labor as it did on the availability of hydroelectric power.[39]

Although there had been a loose ethnic segregation of the labor market before the war, these boundaries hardened after 1915, and the workplace became "an increasingly hostile place for. . .workers from what were now enemy countries."[40] Anti-German and eastern European sentiments in Ontario mining camps have already been thoroughly documented by labor historians, and so will not be explored at length here.[41] It is worth noting, however, that racial divisions of labor in mining were not new. During the gold rushes in the United States, Australia, British Columbia, New Zealand, and South Africa, immigrant, black, and Indigenous miners often worked the tailings of white prospectors, filled "unskilled" labor positions in early mining companies, and conducted other forms of undesirable work at mining camps. These structures of work were enabled and reinforced by larger structures of racism and discrimination in mining societies. The introduction of large-scale industrial machinery after the turn of the century and the wartime creation of new classes of marginalized workers after 1914 were a continuation of a long-standing tradition.[42] Escalating scales of production created bigger spaces for unskilled labor in the global gold mines that an influx of non-Allied ethnic groups helped fill during wartime. Fear of foreigners, particularly Austrians and Germans, and a hatred for "the red element" meant that mines became locations of deep internal distrust. There was little sympathy for strikers, workers were fired frequently without notice, and companies kept lists of undesirable and blacklisted individuals throughout the war years and well into the 1920s.[43]

This internal distrust created unequal consequences for Porcupine miners and residents. In addition to bearing the open dislike of coworkers and mine management, "aliens" in Porcupine became concentrated in parts of the mines where they experienced disproportionate risks to their lives and their health. The connection between ethnic labor and increased risk is apparent in the *Porcupine Advance* headline quoted in this section's subtitle: "3 Accidents in Camp Last Week Russian

Blown to Pieces; Greek Buried Alive in Gravel Pit, Italian Crushed by Rock." Porcupine's gold mines were characterized by multiple vertical levels that required miners to travel long distances underground to go to work. The verticality of the mines increased the risks of so-called gravity accidents, whereby miners or rocks fell unexpectedly long distances or stopes collapsed. In the warm, dark tunnels, miners often worked hard near large, powerful machinery, including drills, blasting equipment, and hoists. Proximity to explosives carried a constant risk. Dynamite had to be packed deep into the rock and then lit from a distance, but unexploded pieces left in drill holes proved deadly. Statistics tell the larger story. Over the course of the war, Porcupine's Italian population doubled, its Chinese population more than tripled, and Austrian, Finnish, Polish, and other eastern European populations increased significantly.[44] Accident rates show that non-English-speaking workers usually made up most victims in mines: over the course of the war, 110 of 199 fatalities were non-English-speaking workers.[45] Rather than address the issue, the Ontario government typically blamed the victims. They claimed that foreign workers were ignorant and "slow in adapting themselves to their work and surroundings. . .and are as helpless as children in protecting themselves from injury."[46] The statistical association between non-English-speaking workers and accidents did not diminish over the years and held true even in 1915—the year when most "alien" workers were (temporarily) fired and replaced with workers from trusted Allied nations.[47]

The Allies feared Germany's perceived power in the international mining economy, and events at Porcupine are part of larger efforts to restrict the ability of "aliens" take out licenses, mine, or process metal in multiple Allied states.[48] Ontario administrators, including Thomas Gibson at the Ministry of Mines in Ontario, collaborated with their federal counterparts and the British government to institute restrictive provisions mirroring those already in place in England, Australia, and the United States.[49] Ontario's mining administrators saw themselves as an active part of a broader global effort to control the Axis threat to the world's mineral resources.

The ethnic division of labor became Porcupine's largest inheritance from the war. The inequalities in the distribution of risk continued and even long after the conflict ended, even as the scale of mining picked up in a recovering postwar economy. In 1916 the British Royal Commission on Natural Resources, Trade, and Legislation in Canada said of

Porcupine that "for common labor and shovelers, Italians and Russians are mostly employed; underground drilling mostly done by Finlanders, Swedes, and Austrian Poles. Canadians, English, Irish and Scotch are employed as mechanics, wood workers, in the mills and other surface operations, while the engineers' staffs are practically all Canadian."[50] In 1920 an editorial in the *Northern Miner* argued that the government should bring in Italian laborers to do "the heavier kinds of mining work." In a private letter from mining engineer Howard Poillon to Dome vice president and general manager Henry Depencier, Poillon opined that the mine might fill its labor needs with hardworking Cornishmen, long appreciated for their mining experience and identifiably English, while French Canadians and Finns were unreliable.[51] By 1924 a student geologist working at McIntyre Mines observed that "laborers are mostly foreigners including Russians, Italians, Swedes, Finlanders, Croatians and many others. The timber-man, track-men, surface men and cage men are mostly French Canadians. . . . Practically all the drilling, blasting and mucking is done by foreigners. Sinking and raising shafts, and the shop work is done mostly by Canadians."[52] When the war ended, the patterns of labor it had instigated would remain in place. As a result, the postwar boom that would follow came at the cost of the health and well-being of racialized workers filling the most difficult and dangerous positions in the big mines.

Professional Geology and the "Disappearing" Prospector

Management of "alien" labor was not the only front on which the Canadians collaborated with wartime allies.[53] Porcupine also began experimenting with international technologies, including diamond drilling. Before the war, some Porcupine engineers and miners disparaged diamond drilling as expensive and unreliable.[54] First brought to Ontario by the American owners of the Silver Islet mine in 1873, it had proved finicky for Canadian geological surveyors and exploration companies.[55] Although it had been used effectively in hard-rock camps of South Africa, miners and managers were still unsure about its merits when they debated the subject openly at a meeting of the Canadian Mining Institute in 1912.[56] The shortage of men and materials during the war made this low-labor, low-input form of exploration more appealing. It quickly proved successful, to the extent that the *Porcupine Advance* observed in 1916 that "Porcupine [was] honeycombed by diamond drills."[57] The extensive diamond drilling conducted during the war added significantly to geological understandings of what lay beneath the rock in Porcupine, even if the mines did not have the people or the equipment to exploit it. Diamond drilling and core-sample analysis on Dome Extension claims helped the company discover the extent of its main ore body underground.[58]

International science, including the growing field of geological science, had long played a role in Canada's understandings of its own resources. In the words of Queen's Geology professor W. B. Baker, the 1909 discovery of the Porcupine deposits "brought the world's greatest geologists to Ontario, the pre-Cambrian rocks naturally came in for the most detailed study and discussion."[59] This work accelerated during the war. In 1914, C. P. Berkey, professor of petrography at Columbia, visited the Porcupine Crown mine to study the rocks there.[60] In 1915, A. G. Burrows wrote about the relationship between quartz and granite in Porcupine by referring to a report by C. R. Van Hise on the Black Hills in South Dakota, another by J. E. Spurr on the Yukon, and De Launay's "The Worlds Gold."[61]

The exchange of geological knowledge went two ways. In his discussion of the temperature of formation, Burrows cited Lindgren's "Mineral Deposits," which associated Ontario's gold-bearing deposits with high-temperature deposits found elsewhere in the world. Burrows claimed Hollinger's rock challenged Lindgren's theories, because the mine's ores showed that gold could be deposited at intermediate temperatures.[62] Porcupine would go on to become a kind of baseline for building understandings of surrounding geology. In Boston Creek in 1916, the annual report noted that the porphyry dikes "resemble the quartz porphyry at Porcupine." In the description of the "Goodfish Lake Gold Area," Burrows and Hopkins stated that "this rock is identical to the rock that shows large 'eyes' or quartz and occurs with the pillow lava flows in Porcupine."[63] Geologists built understandings of new mining areas by directly applying knowledge from other comparable areas.

Increased drilling combined with geological study bore tangible results. In 1915, on the Porcupine Crown, mining geologist A. R. Whitmen successfully mapped a broken vein that had historically mystified experts. He did the same for McIntyre, where "the location of the ore bodies has been rendered difficult by the presence of compression faults which have displaced portions of the ore."[64] With production on hold, the mines were at least able to gain a better grasp of the complexities of their rock and create additional certainty about the location and profitability of Canadian Shield gold.

Yet geologic science informed by international experts could not replace the place-based experience of local people. Governments and mine owners worried deeply about the loss of local mining knowledge (specifically, the old-time prospector) in a globally tied industry guided largely by foreign-trained experts. Concern about dropping prospector numbers first appeared during the war, when numbers would have been low already due to war recruitment.

As Thomas Gibson wrote in his statistical review for the 1918 annual report of the Bureau of Mines, "Gold and silver mines are not like farms, and cannot be worked for ever [*sic*], or even for many years. There must be a constant succession of new properties to take the place of those being exhausted, otherwise the industry will languish."[65] Here Gibson articulated the nonrenewable nature of mining, a faith in northern wilderness to supply new mines, and, most important, the necessity of prospectors to discover those mines. Similarly, the *Northern Miner* saw prospectors as providing an essential function in "opening up" unexplored land in the service of the Canadian economy. This was a project that gained urgency during the war, when mineral supplies and finances were short.[66] Enterprising individuals, often local people who knew the land, were the ones who had enabled the "first discoveries" that the government and the industry credited for Porcupine's origin. International science could never replace those people. As the specter of exhaustion loomed under increasingly difficult economic conditions, Gibson and his contemporaries worried about what would come after Porcupine was done.

Gibson and the *Northern Miner* may have had good reason to worry. Prospecting, with its attendant trenching, assaying, and sampling, slowed significantly during the war.[67] Mobile men in their working years were exactly the sort most likely to be recruited.[68] Besides, there was no money available for risky exploration ventures. Prospecting license sales dropped abruptly. New areas, including Shining Tree, which experienced a minor rush in 1915, could not sustain development.[69] The same was true of the short-lived Kow Kash gold rush in the same year. Had industrial mining killed the prospector? The *Northern Miner* lamented how the North's hardy independent prospectors had been "reduced to the eight-hour day and the regular pay envelope in mine and mill."[70] These men might temporarily reenter the bush at the news of the new rush—prospectors flooded into Ontario from Winnipeg, New York, and "Iron Country" in Michigan for Shining Tree and Kow Kash.[71] But excitement quickly shriveled under the risk-intolerant pressures of the war economy.

Anxieties about the disappearing prospector overlooked the way that the role of local knowledge had transformed since 1909. Enterprising individuals doing different types of work were essential to the survival of existing mines during the war. For example, on the Dome Extension property, a man named H. C. Anchor worked steadily through 1915 until at least 1918 on small-scale, independent daily activities like assaying, planning infrastructure, and collaborating with neighbors.[72] On one occasion he borrowed the geologist ("Mr. Kraft") from Dome to come help him assay.[73] Anchor worked mostly alone and aboveground, until snow covered the claim. In the spring he hired "two white men" to help with the trenching and blasting,[74] and he "borrowed a man from Dome Lake for a little while to do a little digging and draining."[75] This sort of surface work was not unlike the exploratory work a prospector might do on his own claim, the only difference being that Anchor was supported by the infrastructure of an existing company. His work kept the company alive during the war years by fulfilling development requirements and providing tangible results for investors.

But Anchor and his contemporaries did not look like the ideal of the old-time prospector, so those concerned about prospector numbers lamented their perceived loss and expressed anxiety about the fresh-faced mining engineers unleashed on the North from the new mining schools. In the conservatism of the slowdown, administrators worried that a certain well-rounded holistic knowledge of the land would be lost to the book-learned expertise of professional geologists. Southern-educated professionals lacked practical skills and local know-how that made them ineffective in the northern camps. The *Northern Miner* disparaged the shiny government-produced prospector guides, prospector workshops, and soft southern mining-school education.[76] The argument went that good-quality mining men got their education from the land, not governments or schools. The paper lamented the loss of those who lived "the life that is near nature and to the rainbow dream of the future."[77] In the practically focused war years, mining seemed to have taken an unwelcome turn, lacking both the effectiveness and the romance of the old days. To the *Northern Miner*, this was not just sad but economically threatening. The problem received media attention in 1915, when an American engineer came to Porcupine and wrote "one of the most extraordinarily inaccurate reports that it has ever been our painful duty to pursue" (according to the *Canadian Mining Journal*). The man had sampled the wrong parts of the rock for gold (the country rock rather than the schists) and concluded it contained no gold. The *Porcupine Advance* argued that "unique conditions at certain properties" meant that "even experts might be wrong" without the help of local knowledge.[78]

Yet when mining managers and bureaucrats actually encountered local knowledge in Ontario, they complained about its shortcomings. The lack of scientific education could make real live prospectors difficult to work with. The ideal solution, according to the *Northern Miner*, was that the "unscientific instinct" that had served prospectors so well in the past be honed through scientific geology.[79] Others echoed this sentiment. In the spring of 1917, A. R. Whitman, a respected consulting geologist in Ontario, explained to the Cobalt Board of Trade that many old-fashioned mining men saw geology as "too theoretical for use." In order to actually help men find gold, "a knowledge of the earth must be useful." Geologists should strive for usefulness because old-fashioned mining men remained stuck in their old ways, and "dogmatism is the great stumbling block for science." Science was truth for Whitmen, and thus it was essential that the old-style mining men buy into it. Whitmen believed that there remained large undiscovered reserves of gold in Canada's North, and so the "conservative attitude" of the modern mining industry was "injurious to. . .proper development."[80] The risk-taking independent prospector was a necessary predecessor of successful, stable industry—but only if he could make practical use of science.

In a column giving advice to young mining students (engineers and geologists) in 1920, the *Northern Miner* recommended taking on a variety of small jobs so that, once the perfect job came along, the student would have developed a multifaceted approach to mining. "An expert along any one line is like a diamond with but one bright face," the paper chided.[81] One way of ensuring multifaceted diamonds was to incorporate practical experience into mining education. This mandate had been written right into the foundational documents of the new Queen's School of Mining and Agriculture in the late 1890s. Part of a cohort of new practically focused schools of mines cropping up in the United States, South Africa, and Australia, Queen's would teach "the application of Science to mining, the training of prospectors and the economic treatment of minerals." Specifically, "students would be trained in the winter months in Practical Chemistry, blow-piping, assaying, methods of mining, the composition and relative position of rocks and minerals, and in the summer, they would go out to prospect and to determine accurately and scientifically the mineral wealth of counties in detail." Winter instruction should directly enable students "to apply the lessons learned on returning to their summer's work."[82] This directive was carried out, and the schools developed working relationships with Porcupine's mines for the purpose of sending students into the North.[83] Students gained experience working in the mines in a variety of areas around the property. Students were fully immersed in mine life: one described what happened when an ore shoot gets "block-holed" and must be worked loose with a crowbar or explosives, suggesting that he was not spared the more difficult and dangerous parts of the work.[84]

The bureau expected prospecting to pick up again after the war, when soldiers and capital returned from Europe, but by 1920 it still had not made a comeback.[85] The *Northern Miner* blamed the government. Timber interests and the high costs associated with patenting claims were to blame for keeping enterprising men out of the woods, the paper lamented. Cutting red tape and expanding the railway system was what the profession really needed. The periodic government prospecting schools run at Porcupine and other northern communities were scoffed at as a conciliatory gesture in the face of a need for real tangible action.[86]

If the romanticized self-made prospectors of the good old days of mining exploration no longer graced the Abitibi by 1920, locally derived practical knowledge of prospecting persisted. The success of the Bureau of Mines' prospecting classes is a final indication of this. The Bureau of Mines expanded its educational classes in 1920 due to popular demand when classes at Porcupine filled up rapidly. The two-week courses were led by experienced government geologists who taught mineral identification, geology, and blow-piping (where labs were available).[87] The *Porcupine Advance* described attendees as mostly "old-time prospectors," but also students and a few experienced mining men hoping to "brush up" on geological knowledge.[88] These were the individuals who remained "on the ground" in Porcupine throughout the industrial period, reading and responding to the land and its geology with a combination of personal experience and hard science. Not only did these individuals persist on the land, but they also enabled industrial development and expansion on the land during periods when external factors threatened the continued viability of mining in the Abitibi.[89]

The Urgency of Mining on the Home Front

Mining development (under the guidance of locally knowledgeable professionals) carried additional urgency in the context of deep insecurities about the North's productive value during the war. Elsewhere in Canada, Canadians directly equated food production, rationing, and agricultural self-sufficiency with patriotism.[90] But agriculture had never been particularly successful in northern Ontario despite the efforts of boosters, and its failures became more painful during the First World War.[91] Porcupine's residents and stakeholders were uncomfortably reminded of the Bureau of Mines' promises of an agricultural frontier in the North—promises that had been set aside in the excitement of the gold rush, but gained new life under "Home Front" initiatives. Ultimately, the failure of agriculture in northern Ontario would lend mining new rhetorical urgency during the war.

In 1915, despite a half decade of mining, the agricultural dream persisted. During wartime supply shortages, A. G. Burrows of the Bureau of Mines dusted off and paraded a host of optimistic agricultural predictions not seen since the Parks, Kerr, and McMillan surveys. He reminded his audience that Porcupine lay south of the Manitoba border with the United States and just south of the Ontario clay belt. He argued that "it is desirable to have a great part of the townships immediately north of the Porcupine gold area, which are in the clay belt, settled, as the farmers will have a near-at-hand market for their produce" in the mines.[92] In Burrows's view, if land in the North was managed correctly it would soon sprout with agricultural crops. There was certainly sufficient demand. The mines regularly imported produce by train from the South and actively sought ways to source food locally.[93]

Despite Burrows's optimism, farms failed to materialize. In 1916 the Ontario government began offering limited support to settlers who would come to live in the North.[94] When these incentives proved insufficient, the provincial government took things a step further by simply offering free lots to farmers who would work the land in 1917. When this, too, failed, the government proposed an agricultural high school for training youth in northern farm techniques.[95] The Northern Monteith Government Farm ran crop experiments (including potato experiments in 1916) and recommended residents keep bees for pollination.[96] Later it recommended hardy varieties of northern flax and turnips.[97] However, after years of lukewarm results, it gave up even on those crops and turned to stock raising.[98] No amount of effort seemed capable of bringing about the agricultural dream. Resigned to northern Ontario's apparent failure, the *Porcupine Advance* enjoined residents to keep trying because "every little helps."[99] The paper celebrated the meager accomplishments of those who did manage to bring in plants and described how the women of the district met to discuss food conservation and rationing.[100]

As the war dragged on, the notion that the North might not be pulling its own weight on the home front provoked defensive reactions. Arthur Cole, president of the Canadian Mining Institute and mining engineer for the T&NO, reminded listeners at a public talk that mining had done a lot of good for Canada. Although the T&NO had originally been intended for agriculture, ore made up the bulk of traffic. Cole claimed that agriculture would not have provided sufficient business to run trains as regularly or profitably as they did under the mines. Then he explicitly linked the metal industry to the Allies' ability to win the war and criticized "the public" for underappreciating the essential role of the mining industry supporting the Canadian economy in wartime.[101] His speech highlighted northern mining's invisibility in comparison to the settled and productive South while asserting the industry's legitimacy and importance to the success of the state. Porcupine was no great agricultural hinterland, but it still contributed to national projects in critical ways. In the absence of demonstrable agricultural production, the success of the fledging mining industry gained additional urgency.

With a better idea of the layout of their deposits, the mines eagerly anticipated the end of the war. But the return of "normal" conditions took much longer than hoped.[102] Following the November 1918 Armistice, the *Northern Miner* predicted reopened mines, higher profits, and skyrocketing success.[103] Yet the price of supplies remained high into 1920. The annual report of the Dome Mining Company gloomily confessed that "contrary to our hopes and expectations, costs in mining. . .have continued high since the end of the war."[104] Strikes continued to plague the mining camps at both Cobalt and Porcupine.[105] Even as men began to pour back into the camps in 1919, there was not enough to satisfy demand, and returning soldiers expected good pay and benefits.[106] As mining engineer Howard Poillon complained to Dome vice president and general manager Henry Depencier in 1920, "It seemed like the hand of fate, that after having almost within our grasp a satisfactory supply of efficient labor it was removed."[107]

The mines partially blamed the "handicaps" of regulation for the slowness of their recovery.[108] The cry for deregulation hailed back to gold-rush traditions of free enterprise glorified in California and Australia—but on a much bigger scale. As the voice of industry, the *Northern Miner* argued that Porcupine's quick recovery made it the world's "only flourishing gold field" in 1920 and called on Canadians to support their northern mines by cutting red tape.[109] Rhetorically, the mines had a moral duty to provide the world with gold and speed postwar economic recovery at home and abroad.

Despite complaints, the events of the previous decade had set the Canadian mines up to take their place on the international stage. The global demand for metals combined with slow recovery in other parts of the world to put Porcupine mines in an advantageous economic position.[110] If expansion was modest, it was also steady. Dome purchased the stagnant Foley O'Brien mine, and Hollinger bought Schumacher in 1922, while both upped production on existing claims.[111] With improvements to geological science, the return of labor, and stabilizing global economic conditions, the companies once again felt comfortable extending their reach.

Expansion of all kinds inevitably required additional power. During the war, power demands had remained roughly within the limits of locally available sources. The mines traditionally depended on a combination of wood (in its earliest days), coal-fired steam plants, and hydroelectricity. But wood simply could not meet the massive needs of the industrial mines, and labor difficulties among American coal suppliers made coal an unreliable fuel.[112]

Hydroelectricity increasingly appealed but had not been without its own historic difficulties. Although rivers abounded near the mines, there never seemed to be enough power to go around. As early as 1914, the *Canadian Mining Journal* discussed the possibility that Hollinger would be restricted by its waterpower supply. In 1915 the *Northern Miner* observed that "practically all the principal mines of the province are driven by hydro-electric energy." That year, W. R. Roger's statistical review for the Bureau of Mines observed that "of late years there have been clashes between lumbering and water power interests, the former naturally wishing to take advantage of the spring run-off for floating logs, and the latter to store it for use during dry periods." Amendments made to the Rivers and Streams Act of 1915 gave authority to the minister of lands, forests, and mines to adjudicate such conflicts. But Porcupine's power problems would persist because the existence of an adjudicating body to divvy up water resources could not solve the basic problem of supply. In 1916 the *Northern Miner* observed that "more power is needed in Porcupine to keep up with the rapidly increasing demand." It explained efforts by the Northern Ontario Light and Power Company to put in two new turbine units and wood-stave flume at Wawaitin Falls on the Mattagami River with the capacity to produce an additional 3,000 horsepower. Although the first plant had been expected to be more than sufficient for the Porcupine camp's long-term needs, by the end of the war its capacity seemed laughably insufficient.[113]

The essential problem had always been that, no matter how much capacity they added, the plants could not compensate for the massive fluctuations in water level inherent in northern Ontario rivers. February and March had become times of regular power failure because of unpredictable thaws and snowpack in the North. The Bureau of Mines pointed out that such fluctuations might be managed with better storage. If only large-enough reservoirs could be built to hold water, dry periods could be mitigated by slowly releasing buildup from wetter months. The problem was voiced as one of inefficient management of the environment. "Storage facilities. . .are very meagre, and consequently the river cannot be described as well-regulated in its natural condition," opined the *Northern Miner* during a period of low water levels.[114] With better management, the environment could be made to produce the power the mines needed. The kind of large-scale, basin-wide management required was not forthcoming. In the spring of 1917, problems caused by runoff at the Sandy Falls power plant forced the mines to temporarily shut down.[115] That spring the Miller Lake dam broke and flooded school buildings in Porcupine.[116] It was repaired, but only two weeks later it flooded again.[117] During low water periods, the mines were forced to slow or stop production. The extractive capacity of the mining companies, fueled by the global market economy, had come up hard against the reality of an environment that could not, in its current organization, provide the power it needed.

This was the context in which Hollinger sued Northern Canada Power in 1921. The two entities had a complicated history: the Canadian Finance Company had owned both before 1912. At the time, financing a power company had been part of its strategy to ensure the success of its mine holdings. When the two separated in 1912, Northern Canada Power and Hollinger signed a contract guaranteeing that the Hollinger would continue to buy power from Northern Canada Power and that the power company in turn would supply enough electricity to run Hollinger's operations. At the time, Hollinger's power requirements had been around 700 horsepower. This was easily supplied. Over time, both companies changed. Hollinger became Hollinger Consolidated with the addition of Acme and Millerton in 1914. Northern Canada Power had also grown, taking on additional contracts with other mining companies and increasing the capacities of its plants and storage facilities to meet the rising demand. Both parties nevertheless continued to fulfill their obligations.

Then, in 1920, Hollinger claimed it needed 10,000 horsepower.[118] On paper, the capacity of Northern Canada Power's plants actually should have been more than enough to supply this exponentially larger amount, but the extreme low water over the winter of 1919 to 1920 defied on-paper capacity.[119] Hollinger's suit for damages in 1921 ended unsuccessfully at first. The trial judge ruled that, although Northern Canada Power had agreed to supply power to the original Hollinger, Ltd., this agreement did not extend to the greatly expanded Hollinger Consolidated. Furthermore, Northern Canada Power could not be blamed for the low water levels of 1920, which were due to natural conditions beyond the company's control. Although likely a (temporary) relief to Northern Canada Power, the suit solved few problems. Northern Canada Power was under considerable pressure to expand, and not just from the Hollinger. At the end of 1921, the *Northern Miner* proclaimed, "Power Supply at Porcupine Must Be Increased." Now operating at full postwar capacity, the mines overloaded the existing resources of the power companies on a regular basis.[120] Then the Ontario Court of Appeal reversed the earlier decision in Northern Canada Power's favor. The fact that the local watershed simply could not provide enough power to sustain low-grade industrial mining now carried little weight. With so much human and financial capital already sunk deeply into Porcupine's earth, and so much potential wealth still trapped in the country rock, northern Canada *owed* its clients sufficient power to exploit it.

Although Hollinger insinuated otherwise, Northern Canada Power had not been negligent in its efforts to solve northern Ontario's chronically unpredictable water problems. By the time the 1919 shortages hit, it had already been working hard to secure a larger reservoir for water. At around the same time as Hollinger and Northern Canada Power lawyers faced each other in court in the spring of 1921, general manager of Northern Canada Power J. H. Black wrote to the federal Department of Indian Affairs to negotiate a settlement. He wanted to build a new dam that would flood some of the timber on the Mattagami River Indian Reserve.

Black wanted Indian Affairs to send a representative to examine the area and suggest an arrangement for damages.[121]

The relationships between local knowledge and industrial expansion in the North played out particularly clearly in the events that followed. Indian Affairs asked local agent Henry J. Bury to report on the potential values that might be lost by flooding the Mattagami Valley. Bury replied that the land had no value, the timber being small and the alternately rocky and swampy soil being no good for agriculture. However, he also noted that the valley was home to three houses and three gardens owned by Chief James Naveau and his relatives Thomas Naveau, William Naveau, and Jimmy Naveau Jr. There was also a cemetery where "60 Indians and 4 whites" were buried next to the old Hudson's Bay Company post and adjacent to the reserve. Under the circumstances, he recommended that Northern Canada Power compensate the people living there for their houses and "also be responsible for the entire cost of removing the Indian cemetery to the new location."[122]

Bury's assessment differed wildly from that of the Mattagami who called the valley home. Chief William Naveau and his family asserted that Indian Affairs had underestimated the damage and worried that there was no satisfactory alternative location for them to live. "We do not care to live away back in the bush where we could have to go if the Power Company raises the water," wrote Naveau in 1921. The department wrote back to reiterate its earlier offer and state that it was not considering "furnishing additional lands in lieu of those which will be flooded."[123] This was hardly a satisfactory answer. Bury's assessment, which fitted existing expectations for the North as well as its structures of social and economic power, informed and facilitated the expansion of industrial capitalism. Naveau's was rejected and ignored.

Historians such as Jean Manore and Brittany Luby have written extensively about the exchange between Anishinaabe people, power companies, and the government, linking this history to the broader history of colonialism in Canada.[124] Indeed, that the Department of Indian Affairs, the Ontario government, and the mining and power companies would prioritize industrial development over Indigenous livelihoods is not surprising considering the context of relations between the government and First Nations at the time. In 1919 the Department of Indian Affairs had overseen changes to the Indian Act that would facilitate exploration and mineral development on reserves.[125] Relaxed regulations immediately resulted in increased pressure on Indigenous land by prospectors and mining companies, who entered reserves and successfully obtained licenses to stake claims in the early 1920s.[126] This was a change from the less permissive prewar years and is indicative of both the industry's growing lobby power and Canada's mounting contempt for its 1905 treaty obligations.[127]

The story is one of deep injustice in which the Canadian government, the Province of Ontario, and the mining industry worked together to secure Mattagami land for hydroelectric development. Mattagami people actively rejected the project, to the extent that William Naveau refused his compensation check in protest when Northern Canada Power tried to deliver it to him in 1921.[128] There were some additions to the amount of compensation offered throughout the negotiations, but when flooding commenced it became clear that there had been significant oversights.[129] Additional properties were damaged—Alex Pahguish of the Whitefish Reserve claimed that the water surrounded his house and flooded storage areas, damaging his goods. Pahguish, who possessed more resources than Naveau, went straight to his lawyer.[130] His destroyed property was assessed at $850, but in September 1923 the Department of Indian Affairs decided he would be compensated only $500.[131] Northern Canada Power countered with an offer of $200.[132] Pahguish did not actually receive his money until the summer of 1925, three full years after his home was destroyed by the power company.[133] The dam also flooded more timber than had been originally anticipated. The Department of Indian Affairs did not compensate Mattagami for this additional land until the rest of the reserve was flooded (for further reservoir expansion) after 1930.[134]

Hollinger's suit against Northern Canada Power ended in 1925 with an out-of-court settlement.[135] According to mining-media rumors, Northern Canada Power guaranteed the mining company lower rates and other benefits in exchange for dropping the suit.[136] Neither the end of the court case nor the completion of the Mattagami reservoir solved the Porcupine's power problem. Well after the reservoir became operational, member of Provincial Parliament Mac Lang railed against the slow action of Premier Drury's "Farmer Government" on the matter of water power in 1923: "The shortage of electric power. . .is seriously handicapping the operation of the mines," he argued. In response, Attorney General William E. Raney "hazarded the opinion that the present shortage was due to the severity of the weather." To this Lang scoffed that "the weather was not something which was here today and gone tomorrow."[137]

The exchange almost exactly mirrors the arguments put forward in 1921 and is indicative of the hollowness of the assertion that northern Ontario's power problems just wanted for better environmental management.

Hollinger versus Northern Canada Power and the story of the Mattagami reserve showed that northern Ontario's water had a central role to play in the industrial plans and aspirations of northern Ontario. Despite the apparent limits of power, there was still no question that the landscape could support extractive expansion in the minds of the industry's supporters at local, provincial, and international levels. After all, the Bureau of Mines had been publishing long lists of potential waterfalls suitable for hydroelectricity plants since William Parks first traveled through the area in 1898. Doubt about the future of mining was unthinkable against a long history of industrial optimism, especially when that optimism was backed by enormous human and capitalist investments already sunk into Porcupine rock from around the world.

The embarrassing wartime agricultural failure made mining's success even more intellectually imperative, as proof of the North's ability to contribute to the global economy. Unlike agriculture, industrial mining had a future at Porcupine. The shield's hard-rock deposits, the Great Fire of 1911, the war, and finally by a roaring 1920s investment economy all added up to the success of mining in the North. If the landscape would not oblige, it was the duty of the government and the mining companies to manage it. The costs of doing so, in the case of the power crisis, were offloaded to people already marginalized by Canada's colonial government, whose interests came second to ensuring the health and continuity of gold mining.

Roaring Markets, Silent Rivers, and the Aftermath of the War Years

After 1922 the power problem seemingly solved, an overwhelming sentiment of optimism permeated Porcupine mining. The *Porcupine Advance* and the *Northern Miner* began publishing triumphant accounts of Porcupine's progress since 1909. These pieces included a brief description of the region's history and its founding "discoverers," the latest news from the mines, and a series of incredible statistics; one of the earliest boasted that "during the past ten years, over eighty-five millions [*sic*] have been taken from. . .the Hollinger, the McIntyre and the Dome" and predicted that these amounts barely "scratched the surface" of future possibilities.[138] The federal and provincial governments commissioned films on the mining industry featuring Porcupine, further integrating it into broader national stories.[139] Porcupine became part of a larger story of national advancement, improvement, and industrial triumph in 1920s Canada.

These stories of triumph belied nearly a decade of crisis and insecurity. As the war sapped labor and supplies, the mines faced the possibility of the end of profitable extraction for the first time. These anxieties were amplified under "home front" rhetoric that celebrated agricultural accomplishments that proved impossible in Porcupine. The industry and the government worked together to frame mining as a national project equally valuable to that of agriculture and to develop a cadre of professional prospectors who would ensure the continuation of mining's legacy. The end of the war should have brought relief from these conditions. Yet without adequate power resources, the mining companies' goals remained frustratingly out of reach. In 1920, with reservoirs empty and the mills shut down, the dream of unfettered industrial extraction on Porcupine's mining frontier might have remained just out of reach. Instead, mining companies, power companies, Indian Affairs (federal), and the Bureau of Mines (provincial) worked together to reorganize the landscape in a way that would allow them to pull additional power out of northern rivers and push the mines to new extractive scales.

The industry's new power source came at a social and economic price to Mattagami people who were inadequately compensated for serious losses to land and resources. At the same time, the First World War created a cheap workforce of racialized migrants whose bodies worked in concert with the machines to power the mines while alleviating historic labor shortages. Such inequalities made up the bedrock of human relationships with nature at the Porcupine through the rest of the 1920s. In the aftermath of the Great War, Porcupine's mining industry overcame economic, human, and environmental challenges to become one of the world's greatest mining areas.

CHAPTER FOUR

Mine Waste

Environmental Disaster Above- and Underground

Although the celebratory narratives of the early 1920s pitched Porcupine as an extension of a romanticized frontier of California and the Klondike, the low grades and high costs of extraction made mine ownership and development unlikely for individual prospectors.[1] In the ten years that followed, these conditions reached new extremes. Big mines employed hundreds of men who worked specialized jobs. The large majority were "muckers," who worked to move ore from the underground faces to the ore cars and then from the ore cars through the mills. Meanwhile, mine owners and investors retreated to downtown offices in Toronto and New York. History had favored companies backed by big capital operating under economies of scale.[2]

We have already seen how the gap between the expectations for the Canadian Shield and the nature of its *actual* geology and ecology created anxiety about the future of mining and conflicts over waterpower in the 1920s. This gap manifested in other ways as well. Specifically, the 1920s saw renewed conflict between workers versus management and a nascent environmental movement that put the community interests at odds with the industry. These new tensions stemmed from a series of waste-management issues and resulted in escalating environmental problems that punctuated the 1920s. The provincial government and the mining companies framed these disasters as "natural" and inevitable products of life in the North. In doing so, they obscured the ways that human choices and actions created these events over time and exacerbated their impacts in the days and years after.

Disaster scholars have pointed out how framing "natural" disasters as acute events normalizes the structures of oppression that generated them in the first place.[3] The mining disasters of the 1920s normalized suffering, allowing the industry to maintain the status quo despite increasing imbalances between the scale of operations and the ability of the environment to absorb its wastes. Porcupine became a laboratory for a world coming to terms with the unexpected, voluminous, and toxic by-products of industrial mining. Lessons learned on the Canadian Shield proved especially compelling because they promised that scientific solutions would make industrial mining compatible with its environment.

Chapter 4 looks behind the mills and into abandoned stopes in search of the by-products of extraction that the mining industry ignored until it could no longer afford to. The "big three" extractors offloaded the costs of their expanded production in various creative ways. Ultimately, smaller companies, individual workers, and the wider community would bear the bulk of the costs. As wastes accumulated above and below the ground, risks flowed downstream in both literal and metaphorical senses. On the surface, waste rock filled up local lakes and depressions and spilled out over the land, including land claimed by small mining companies and land used by residents of Porcupine. Deep underground, empty spaces became reservoirs for the detritus of extraction. In the early 1920s, both the surface and the subsurface reached critical capacity and threatened production for the first time. Above the ground, mining companies were forced to find places for waste that would not impinge on neighboring properties. Below, they would introduce a new collection of regulations for mine safety. In both cases, technological solutions would allow the continuation of production while offloading responsibility for environmental safety onto workers and local residents.

The Nature of Postwar Mine Labor

Mining labor looked different after the war. The working environment included a collection of underground and aboveground spaces, including mills, warehouses, change houses, shafts, stopes, and tunnels. Liza Piper describes how miners and geologists described underground environments like bodies, complete with veins, heads, tails, and faces. This language helped render an alien world familiar by drawing on "familiar organic metaphors." Each mine's headframe, which contained the hoist for the main shaft, became the point of connection between the surface and the underground world.[4] For workers, the underground environment was qualitatively different from the surface, despite the occasional insistence by outsiders of its natural healthfulness.[5] A collection of infrastructures, including ventilation, mine timbers, lamps, and compressed air, were required to render the underground safe and habitable. Even with these infrastructures in place, underground environments remained dangerous, mysterious, and occasionally hostile. Although miners, geologists, and managers worked together to build these environments, they remained unpredictable and only ever partly under human control.[6]

It is important to understand the nature of postwar labor because the minds and bodies of workers always filtered the desires of bureaucrats, scientists, and mine managers. In a practical sense, things did not happen unless they were done by workers. In Porcupine as elsewhere, 1920s mine workers increasingly encountered the land under the structures imposed by industrial capital. However, they often translated the directives of mine management in unpredictable ways. There is no way to draw a straight line between the history of the big mining companies and the environmental history of Porcupine without first acknowledging the agency and complexity of the people whose hands literally enacted environmental changes.

A series of letters between Dome Mining Company vice president and general manager Henry Depencier and one of his employees demonstrates how these relationships operated on the ground and, further, how the practical work of mining connected to its international context. The letters are the product of Dome's work to counteract gold theft in their mines. Gold theft was not uncommon in Porcupine—police reports chronicle a variety of crimes from individual opportunistic thefts to large-scale heists via organized criminal networks.[7] It was known colloquially as "highgrading" at Porcupine and anywhere else where gold was found. In Porcupine's age of industrialism, opportunistic workers subverted their employers by piggybacking on their laboriously constructed infrastructure to cherry-pick their hard-won products. In one particularly profitable example, three men smuggled more than $150,000 in gold into the United States and sold it to the San Francisco Mint between 1917 and 1921. The scale of this particular heist speaks to the cross-border integration between Canadian and American gold economies. In the aftermath of the theft, Depencier hired an undercover agent to investigate highgrading at Dome.[8]

Depencier's agent (who signed his name only as "A. F.") worked throughout many of the mine's major parts in the winter of 1921 and spring of 1922. He worked alongside Dome miners, talked to them, and tried to win their trust. In his letters to his employer, he speculated about the loyalty of specific Dome employees, suggested improvements for mine operations, complained about his supervisors, lamented bad camp food, and related all the latest bunkhouse drama. His writing is surprisingly open and uncensored. He worked during a time when workers were increasingly integrated into complex mining systems requiring large amounts of human labor for generalized functions such as shoveling ore (mucking), cleaning machinery, or construction.[9] They emphasize the extent to which industrial extraction was no unified human effort, but an imperfect collection of competing human interests on the land.

A. F.'s descriptions contradict the idealistic visions of gold mining offered to investors in Dome's annual reports in which extraction was quantifiable and rational. His work was difficult, dangerous, and occasionally frustrating. In one account early on in his time at Dome, management stationed A. F. in the cone room and assigned him to clean an agitator tank that mixed ore with lime. "This is cleaned once every six months, whereas it should be cleaned at least once a week," A. F. reported. As a result, the tank would get choked up with sand, stone, and lime that A. F. had to laboriously clean. He brought the matter to the attention of a supervisor, but the supervisor was dismissive, stating that "the cone men. . .cleaned the tank whenever they had the time." A. F. insisted in his notes that "it is a very unpleasant job and is always put off as long as possible" by those responsible.[10] In another instance, A. F. worked in the mine mill where he became a "slime helper." Here he complained several times about the difficulties "experienced in controlling the slime," which tended to flow unevenly through the mill and needed constant attention. He claimed that it was difficult to manage the slime while also taking care of all the oiling, greasing, repairing, and cleaning necessary to keep the mill running. After bringing the issue to his supervisor, his concerns were again dismissed.[11] The very next day the mixture became so thick and heavy that it clogged the machine and halted work until it could be cleaned out, a process that would take anywhere from ten days to two weeks.

A. F. did not bother to cover his feelings of vindication in his letters to Depencier.[12] He did not understand why the company did not just adjust and slow the amount of slime that was processed. This simple change would avoid overburdening machinery and men. He finally got to the bottom of the matter when he realized that "Frank Horne, the Superintendent, always complains whenever the ball mills have to be stopped." Horne complained because profitability was perceived as directly proportional to the volume of ore the mill processed. When the mill stopped, Horne worried that the shareholders would call the operation inefficient.

However, as A. F. pointed out, while continual operation "may look good in the reports, a ball mill running without stop, it caused a lot of extra work cleaning up unnecessary waste and makes the job of solution helper the most hated job in the mill."[13] Without hands-on experience, investors and managers could not understand such subtle losses to time, resources, and morale. Neither seasonal variability nor miners' annoyance carried much weight in an annual report, and so overloading the ball mill continued at the expense of miners' frustration and labor.

A. F.'s letters also serve as a reminder that the work of mining exposed miners' bodies to great danger.[14] For miners, sickness and injury could have social, economic, and physical consequences. In one instance, A. F. had to work extra hours to cover for a sick coworker: "I had to put in [an]. . .extra 4½ hours because Tom Sawyer, solution helper, whom I relieved, suddenly took sick and had to be removed to the hospital."[15] In trying to uncover discontent among employees, A. F. recorded a long list of miners' worries about their physical safety. On May 26, "I was speaking to a man named Fitzpatrick who is employed underground and he was denouncing the company for not taking better care of the safety of the men working underground, as he said there is continued danger from falling rock." The next day, "Lou Wright, who works underground, said that of all the mines he has worked in, the least precaution for safety of the underground men is taken by the Dome Mines." According to Wright, Dome did not employ a standard "scaling crew" to cut loose hanging rock from expanded tunnels and pits. Wright also noted that the ethnic diversity of miners created an unsafe environment, since foreigners could not understand English warnings and vice versa.[16]

One of the most profound moments in A. F.'s letters occurred on May 27, 1922, when a tunnel ceiling collapsed on the night shift.[17] The event shocked the camp. Miners immediately went to work pulling dead and injured coworkers from the rubble. While working in the power room, A. F. overheard two miners angrily deriding two Polish men who had refused to help with the gruesome task (possibly because the collapsed area remained unstable and dangerous). The two angry men also said that the mine should have stopped work out of common respect for the dead.[18] Management declined to do so, and A. F. (somewhat shaken) worked right to the end of his usual shift.[19] Pulling the bodies of coworkers from mine shafts sparked unease among miners as they confronted the gruesome reality of occupational risk in the industrial mine. There was a sense that the great cog of capitalist extraction should pause for a moment of reflection on the gravity of these risks, which miners subjected themselves to every day. Such sentiments carried little weight for management, for whom the greatest risks lay in expensive interruptions to production. The death of miners (although serious) remained an abstract, unsurprising, and impersonal problem unrelated to the question of whether to shut down the mine.[20]

The richest parts of the A. F. reports describe the ways workers controlled and managed danger where their employers seemed unmotivated to do so. A. F. reported several instances of corner cutting by fellow employees who left work early or sat around on the job. Others purposefully wasted materials in ways that made their job easier. A. F. spoke about mine waste to Archie Dowell, a machine helper, who said "it is the duty of all men to waste as much material as possible, as by so doing he is creating work for fellow workmen employed in factories manufacturing the material."[21] He said that machine men would "go for a fresh supply of grease each shift," even though "only a small quantity of it is used." The extra would be dumped rather than conserved for the next shift: "A great deal is wasted and buried beneath the rock during the blasting. He said the men never try to save it. He said the lubricating oil is also wasted in the same way."[22] In dumping their grease every shift, Dowell and his colleagues rejected the interests of the mine in favor of saving their own labor, wages, and time. A. F. also wrote regularly about miners' dissatisfaction with food. These complaints were serious. Miners pushed constantly for better food and living conditions in camps. During A. F.'s tenure, the state of the kitchens brought employees to the point of strike, forcing Dome to hire new cooks. Miners also had to advocate for freshwater for underground workers. Before that, men working underground were forced "to drink the partly poisonous cyanide seepage water, which is the only water to be had underground at present."[23] Miners objected to the health risks and took actions to correct them, especially when it was forced on them by managers who did not share the same risks.

Miners also managed their personal lives in ways that brought them joy and connection in an industrial world poorly organized for fostering such sensations. Hints of a rich recreational life permeate the sources around Porcupine. A. R. Globe, assistant general manager at Hollinger, observed that in winter, "the cold is not felt, and, as a matter of fact is generally enjoyed" by Porcupine residents. In addition to organized hockey games, locals also enjoyed snowshoeing, dogsledding, and skiing.[24] A sense of camaraderie permeates A. F.'s letters, despite the fact that he saw himself in opposition to his colleagues. The sheer amount and diversity of miners' gossip (ranging from food to mining stocks to venereal disease) suggest an active and interesting social atmosphere.[25] In January, A. F. spent at least one afternoon "skating and playing hockey" with his coworkers, after which he joined them for an evening poker game.[26] His reports of these outings contain tantalizing hints of life on the land enjoyed beyond the confines of work and paint a picture of men who had learned to live in (and with) northern nature. If this large, "unskilled," and somewhat disenfranchised workforce bore the brunt of environmental disasters that rocked the mines in the 1920s, the A. F. reports show how they did so actively and with acute awareness of the inherent risks of their work. The act of extraction was always shaped by their hands, and miners rarely carried out the vision of their managers to exactitude.

Waste Aboveground and the Trouble with Tailings

Part of what A. F. was witnessing at the ground level was how the rapid pace of industrialization and the newness of many of the technologies used at Porcupine in the 1920s created unexpected by-products that were deliberately ignored. This era saw the mines struggling with unprecedented amounts of waste generated during production and little thought about how to manage it. In the past there had been little incentive to develop waste solutions. Instead, engineers focused on bringing ore from the faces through the mill as efficiently as possible in order to maximize profit. After the gold had been taken from the rock, whatever non-gold-bearing material was left over became singularly uninteresting from a mine owner's perspective. Traditionally, mines simply dumped the waste wherever they could find a good spot—usually in a lake or other natural depression. This cost companies nothing. But as the scale of extraction increased, convenient depressions became more difficult to find, and wastes began to spill across the landscape, creating expensive problems for mine management.

Trouble with mine tailings was foreshadowed during the war. Tailings are the fine material left behind after gold has been extracted from ore. Their consistency varies depending on the type of rock being mined. They can be toxic where chemicals are used to separate gold, or where the rock contains heavy metals. At Porcupine tailings waste took the form of slimes—a fine particulate mixed with water. By the spring of 1916, Hollinger's tailings were spilling out of their holding areas to the point that they threatened to block access to its supply road. "The old road has been in danger of flooding by tailings for a year past," reported the *Northern Miner*, "and will certainly be inundated this spring." The company solved the problem by dumping fill between Miller and Gillies Lake and building a new road on top of the fill. The project made Miller Lake into a giant tailings pond, but this solution was considered temporary because projected production would quickly fill these new spaces: "It is not anticipated that it will suffice for more than a year and a half at the rate it is expected that tailings will flow from the mill."[27] Even with this new storage secured, the new road still had to detour around another large section of slimes.[28] On other properties, tailings caused tension between neighboring mining companies. In the winter of 1917, Peterson Lake Company discovered that Nova Scotia and Dominion Reduction companies had dumped three hundred thousand tons of tailings into Peterson Lake. Peterson Lake sued for damages.[29] Waste management was becoming expensive.

A 1923 court case between Digby Vet and Dome Mining Company is a spectacular example of how wastes plagued the mining industry during the postwar era. Tailings from Dome mine had begun spilling onto Digby Vet sometime in 1913. No one was monitoring them, so the exact date is unknown. Dome vice president Henry Depencier later recalled that, at the time, the company policy was to dump tails directly into nearby Edwards Lake.[30] Unbeknownst to the company, the lake filled up, and tailings began to escape through the Edwards Lake's small outlet. This stream flowed across the Digby Vet claim. Unmonitored and unmanaged, the wastes flowed where gravity took them. Occasionally, the tailings would form a temporary dam of thick sludge, diverting the flow randomly until another dam formed. In this way, mine waste built up and flowed over the land in a continuous accumulation.[31]

At some point prior to 1917, Digby management noticed the leak and complained to Dome. Dome's response was notably feeble, but at the very least the company took responsibility for the tailings overflowing from Edwards Lake. It built a sandbag dam across the middle of the old Edwards Lake bed, hoping to keep the tailings away from the outlet. This dam proved "a total failure in the wintertime" and in any case quickly disappeared under the sheer volume of Dome's outflow. Dome also tried a series of elevated troughs and pipes to dispose of tailings by spreading them around the landscape. As new areas filled up, Dome shifted its pipes gradually west to distribute the outflow more evenly.[32] This did not solve the problem, and tailings continued to flow.

Digby Vet, like many small mining claims, had neither the motivation nor the resources to monitor or develop their mining lands during the war. Like many small companies during this time, their hold on the mine was tenuous. Majority shareholder and vice president of Digby Vet William John Aikens, for example, could not describe the basic physical characteristics of his claim, explain the extent of recent work, or even give the correct name for the equipment used to convey mine waste (which he incorrectly called a "sluice"). Aikens was based in Bay City, Michigan, and had visited his Porcupine property only twice—the most recent visit being eight years before his 1923 testimony. The upturn of the mining market and the subsequent intensification of activity changed these circumstances. With increased demands for metals and new sources of available capital, small claim owners, including Digby, began to think seriously about returning to development. It was this change, combined with the increase in Dome's production, that sparked a sudden interest in tailings.

But Digby was not the only entity Dome ran afoul of as a result of tailings overflow in 1922 and 1923. Spring breakups washed tailings downstream where they backed up against the T&NO rail spur. Dome management wrote the railway's chief engineer to request that the line be raised and additional culverts put in to compensate for the changes to the landscape's natural drainage patterns.[33] However, like the Digby tailings, Dome's actions in regard to the railway floods were reactionary stopgaps rather than meaningful solutions. It was clear that the mining company would avoid serious structural change for as long as possible.

Tensions between Digby and Dome escalated quickly. In 1919 Digby's lawyer had written to Dome and formally asked them to remove the tailings from the Digby claim. Dome's lawyer Alex Fasken unhelpfully replied, "As you know it is impossible to move this stuff."[34] As a possible solution in 1921, Digby agreed to consider selling the land to Dome for tailings storage. But they wanted $2,000 per acre, a price Dome described as "so utterly fanciful that it could hardly have been intended to lead to any business."[35] Dome countered with an offer of $50 per acre, and (perhaps understandably) Digby responded with a suit for damages. A breaking point had finally come.

In court in 1923, Depencier's testimony revealed the total lack of thought that went into questions of waste disposal in early-twentieth-century mines. Indeed, if remarkable advancements had been made in other areas of mine technology, little had changed in the way that mines dealt with waste since the nineteenth-century gold rushes. When pushed on the matter of where and when Dome tailings had flowed onto the Digby property, Depencier simply could not answer. He argued that such information was impossible to say for sure "because it changes from day to day to the south" as the slimes moved and settled out from their disposal system. He insisted that even if they wanted to conduct one, a complete survey was impossible while the snow lay on the ground. When Digby's lawyer pushed him, insisting that "this question of disposition of slimes is an important one in any mining venture," Depencier obstinately replied, "I don't understand your question." Digby's lawyer asked how other mining companies did it, and Depencier said that there are "very limited" means available and that he "did not know of any other means in operation today except to discharge them out on to some area." When asked about mining companies outside of Canada, Depencier scoffed that "in a warm climate you can build a dam," but explained that the spring thaw in the North made damming impossible for Porcupine. Digby's lawyer then asked about stacking, which Depencier similarly dismissed: in South Africa he had "seen it tried, but it failed." He then went on to describe how, in all his mining experience in the biggest camps, he had not seen a different solution. British Columbia, Washington, and Michigan mines all used the same methods of dealing with waste. Furthermore, northern Ontario faced some unique challenges beyond what might be experienced elsewhere. Because of the very low-grade and highly dispersed nature of the gold, Porcupine ore must be ground into a fine slurry and mixed with water for processing. Other mines could get away with grinding to dry sand, which could be more easily stacked. Dome could not.[36]

Nevertheless, Digby's suit made it clear that letting Dome's tailings flow unmonitored across the landscape would no longer be a tenable solution to the challenges of waste in a northern environment. Dome would be forced to find a way of controlling their outputs. They needed a place where they could dump their tailings where it would not run into streams and cause potentially expensive damage downstream.[37] At one point they tried to get permission to dump into Porcupine Lake, a major source of water for the communities of Porcupine and Timmins. They were stymied by the provincial government's stipulation that the lake could not be fouled without obtaining rights from all stakeholders in its watershed—a list that included the municipality, which was unlikely to agree.[38] The fact that the company had more or less exhausted its options by 1924 is evident in the fact that, after Digby filed an injunction preventing further dumping, the mine was forced to buy another property (on the other side of Digby) for an exorbitant price and then obtain rights from Digby to pipe the slimes across it.[39]

The Digby suit revealed significant blind spots in the vision for industrial expansion held by Porcupine mines. The immense amount of waste generated by low-grade mining forced mining companies to pay closer attention to what was happening on the ground and to make calculated decisions about how to use the limited resources available to them. Dome's tailings troubles were a symptom of building tensions between those with stakes in the landscape of Porcupine in a context of scant regulatory oversight. They were indicative of the priority given to the production side of mining that provided little opportunity for thinking about its consequences. There was little incentive to think about tailings until Digby's litigation made the costs of ignoring them too high. Mine management, government, and miners would work these out over the following decades with varying levels of success, but to some extent waste management continues to be an unanswered problem for mining companies in the Abitibi and elsewhere. Bitter disputes over mining wastes characterized global mining frontiers before and after Porcupine.[40] While mine "remediation" offers technical solutions to waste accumulation on the land in the twenty-first century, it continues to obscure historic trauma associated with past development by focusing on the benefits of cleanup while overlooking legitimate community concerns and values.[41]

The Hollinger Fire, Waste, and Mine Safety Underground

As the mines struggled to find places to squirrel away their waste, the vast underground spaces they had created through extraction looked increasingly tempting. Indeed, just as mine waste flowed over the land above the industrial mines, it also filled up the spaces underneath. Tailings were not generally returned underground at the turn of the century (later on, they would be mixed with binders and used to backfill old tunnel pits), but underground spaces became receptacles for other kinds of mine garbage. Although waste disposal in unused stopes and tunnels was touted as an economic and low-impact solution to waste accumulation aboveground, it proved just as difficult and costly to manage. Confined to the tunnels and stopes deep within rock, the consequences of waste accumulation could not be avoided. They just took longer to make themselves known.

In 1928, Hollinger became the international poster child of the perils of mismanaging waste when a garbage-filled stope caught fire, sucked oxygen out of active workings, drove flames and gas through shafts at velocity, and filled the rest of the mine with smoke. Miners began evacuating at ten in the morning when they noticed smoke, but with limited hoists, the presence of deadly invisible gas, and no way to tell where the smoke was coming from, retreat quickly became chaotic. The incident killed thirty-nine miners and became the linchpin in a movement toward professional mine rescue and worker safety by gold mines around the world.[42] Gold mines are wet places: the shafts and tunnels at Hollinger dripped all day, and the biggest problems underground stemmed from preventing shafts from filling up with seeping groundwater. Before 1928 gold mine operators considered themselves safe from the explosive calamities that plagued their coal-digging counterparts. After, they realized that waste deposited belowground could be as dangerous and costly as waste deposited aboveground.

The 1928 Hollinger fire occurred at the intersection of managerial alienation from the land and worker agency. As with aboveground tailings, management failed to keep up with the volumes of waste produced underground during postwar expansion. Corporate emphasis on the profitable continuation of industrial mining and lack of interest in the by-products of progress led to serious environmental oversights. At the same time, miners' labor-saving shortcuts made the situation dangerous and ultimately provided the spark. Social and ethnic divides between miners hindered communication and made the disaster worse. The fire violently highlighted a major gap between the promises of northern extraction and the realities of environmental risk under industrial capitalism.

The root cause of the fire was the increased pressures of scale and volume in expanding Porcupine mines. The scale and volume of mining was largely determined by how much rock miners could blast from the working faces during a single shift. By the 1920s, Hollinger routinely stored explosives in naturally warm underground chambers as a cost- and time-saving measure aimed at keeping dynamite thawed and close to blasting faces. The size of the mines and their ability to store dynamite underground were self-reinforcing: deeper mines did a better job of insulating the tunnels from surface temperatures. Before deep mining, dynamite had been stored aboveground in special sheds. Frozen dynamite had to be thawed thoroughly before it could be used, a fuel-expensive (and sometimes dangerous) process. Miners used dynamite daily, so keeping enough explosives warm and ready for use posed a continuous problem in the winter, limiting the amount of rock that could be blasted and therefore the amount of gold a mine produced.

Underground storage solved some of these problems but came with new ones, too. Primarily, this was about waste. For transport and storage, miners stored dynamite carefully in large wooden crates. When dynamite first moved underground, miners onerously moved empty crates to the mine surface and burned them. However, by the 1920s this practice had slowed—partly because its sheer volume of the waste took up space in the carts and hoists that could have been used for men and ore and partly because unused detonators left in the crates sometimes exploded unexpectedly in the garbage fire. Over time, when explosives were unpacked underground, the wood boxes and shavings increasingly got dumped on the stopes along with other underground waste—fuse ends, dirty oil, dysfunctional detonators, rags, and timber ends.[43]

The practice of dumping excessive waste belowground really began in earnest in the early 1920s, with the postwar production boom. It did not take long for huge piles of garbage to build up in underground spaces. In his testimony following the Hollinger fire, safety inspector John Knox stated that the debris found on the burned stope (55-A on the 550-foot level) had been accumulating since 1923.[44] By February 10, 1928, when it went up in smoke, the debris pile was twelve feet wide, one hundred feet long, and forty-five feet deep.[45] As it burned up available fuel in the confined space, it spread smoke and deadly gas throughout the mine. But the pile of garbage in stope 55-A was just one of many similar piles scattered throughout the mine. These piles probably would have been completely safe if the stopes had then been backfilled with a covering of rock, but the informal practice meant that no coordination or planning for backfilling took place and the flammable debris remained exposed.

Before the 1928 Hollinger fire, the mines had little reason to think that their dumping practices constituted any sort of danger. The walls literally dripped with water, and there was very little timber in the hard-rock shafts. Metal mines, as a rule, did not burn in the same way that coal mines did. In the aftermath of the explosion, the *Northern Miner* reflected on the fact that this sense of safety had led to serious regulatory oversight. "The fact appears to be that in no part of the North American Continent where metal mines are in operation is there a specific requirement by law or regulation that empty boxes, paper, and combustible refuse must be brought to the surface and the further fact is that such refuse, as a general practice, is placed underground."[46] The risk had not really occurred to anyone. Like with aboveground waste, there had not been any incentive to manage underground garbage, so there had been no engineering or regulatory solutions. Hollinger had been following the practice of all metal mines in its dumping practices and had not been exceptionally negligent.

Indeed, by 1928, Hollinger was widely considered "one of the greatest gold mines in the world."[47] Its place among the greats made its experience particularly compelling to large international mines facing similar geological, economic, and social circumstances. Porcupine was no longer as remote or isolated as it had once been, and news of the fire spread almost as quickly as the flames themselves. Aid immediately began pouring into the community. Air-quality detection equipment came from Toronto, provided by the Consumers Gas Company, Ltd. More rescue equipment and apparatus, including a rescue car specially designed to enter mines during emergencies, came north from the US Bureau of Mines. "No boundary line was recognized in the act of co-operation between neighbors interested in the same industry," observed the *Northern Miner*.[48] These efforts had tangible effects. The rescue car was so essential in the aftermath of the fire that Noah Timmins wrote to Scott Turner of the American Bureau of Mines on February 21 to thank him personally for its use, writing that "without [the assistance of the car and its crew] we were in a position of helplessness."[49] The railway opened the line, and the equipment arrived in the North in eleven hours, faster than any express train.[50] Similarly to the 1911 fire, news spread rapidly outward from there, making the front page in British Columbia and in California on February 11.[51] Notably, the day after the disaster, several Australian newspapers (in Adelaide and Sydney) had details cabled from Toronto on the cause of the fire and about rescue efforts.[52] By February 13, major papers across New Zealand ran the story, with most of its salient details.[53] By February 15, most papers had an accurate death toll.[54] The papers then followed the process of the investigation into the spring and summer of 1928.[55]

In the immediate aftermath of the fire, the mining companies, government, and mine workers worked hard to identify the cause of this cataclysmic failure. The focus slipped mostly away from waste, which, like Dome's tailings troubles, was seen as more or less unsolvable. Instead of addressing its root cause, the conversation focused on the mine's response to the explosion. In a general meeting held just days after the fire, speakers from Porcupine's diverse community addressed the audience in six different languages. Some expressed "the usual denunciation of the capitalistic system" and used the word "murder" to describe the fire. Others were more hesitant to assign blame, speaking, in the words of the *Porcupine Advance,* in a more "restrained and reasonable" tone.[56] Two inquiries followed, one by the local coroner's jury report and a second by a government commission headed by Ontario Court justice T. E. Godson.[57] The coroner's jury, located in the community and likely controlled or at least sympathetic to the miners, blamed the companies for negligence. Godson's much more forgiving report made unsurprising recommendations around the frequency of mine inspections, explicit rules to cover underground debris, and guidelines for ventilation and escape routes for miners. But many of Hollinger's neighbors did not wait for the results of these assessments. They already knew what had gone wrong. Within a few months, for example, the Creighton Mine instituted a ten-day training course for its employees based on the American Bureau of Mines' "Advanced Training for Recovery Operation During Metal Mine Fires."[58] In August 1928, companies brought in G. W. Grove, an engineer for the US Bureau of Mines to instruct crews building new rescue stations, creating the Ontario Mines Rescue organization.[59] The implicit argument in both measures was that it was disaster response, rather than waste practices, that was at the root of the problem. Changing waste-management practices in the expanding mines was secondary. Explosions would happen. What companies needed to do was to better equip miners to face the new dangers of their deep industrial labor, regardless of its cause.

By emphasizing rescue, the industry and the government addressed the symptoms rather than the problem. By doing so, they shifted the onus of responsibility for the environmental consequences of industrial mining to individual miners and away from the companies themselves.[60] Historians of work and gender have pointed out that miners were international archetypes of working-class strength in unpredictable working environments.[61] This trope could easily be conflated with a broader narrative of human struggle against a hostile northern climate. This was especially true in Canada, where identity formation historically connected with resource extraction and the harshness of the landscape. Such conflation had the benefit of deflecting blame from negligent mining companies to the harsh character of the local landscape. This was especially important in the context of growing concern around health and safety and, in Porcupine, the roots of unionization.[62] Godson's (forgiving) report on the fire made especially heavy use of this rhetorical technique. Godson wrote that "environment molds character" and waxed poetic about the bravery of the Hollinger miners.

> I vividly remember Fred Jackson, quietly and unobtrusively telling his story of conflict with nature's forces. Without exaggeration or boastfulness he recited how he and his four companions retreated from one vantage point to another, slowly and stubbornly backing away from the fumes of the deadly gas; how he turned on the air and directed it against a plank to cause the current to rebound; connected lengths of hose in an attempt to blow the smoke away, and how he cut his smock in four pieces and unselfishly gave his companions a piece to place over their mouths.[63]

In Godson's telling, the miners' battle with rock, air, and fire takes on warlike qualities. Bravery, strategy, and an intimate knowledge of the workings of the tunnels allowed Jackson to come out victorious. Men like Jackson, Godson said, were "a tribute to the manhood of the North, made sturdy and true by their contact with nature's forces and their fellow man."[64] Crucially, their battles were not waged alone but in solidarity with men working in similar conditions around the world who worked together for the common goal of economic and industrial triumph. These narratives overlooked the roles of states and mining companies in producing disasters by making industrial mining (and its consequences) the responsibility of miners.

The publication of formal reports made the lessons of the Hollinger explosion conveniently exportable. Mirroring the content of the official reports, the key takeaways for international audiences drawn from the fire centered on either managerial negligence or mine safety. The accusatory language of the local coroner's jury inquiry seems to have carried the most weight among the international papers. The *Aukland Star*, for example, wrote that "there had been gross negligence on the part of the management and operating executives. They were blamed for permitting rubbish to be dumped in the old stopes."[65] The stories also focused on the fact that men could not escape the fire fast enough because the lifts carried a limited number of people.[66] These kinds of arguments rang true in the context of workers' movements in mining districts around the world and became part of broader arguments for unionization and more stringent measures for worker safety.[67] Any industrial miner could recognize the trope of negligent management and empathize with the terror of being unable to escape up overtaxed shaft hoists.

The papers also emphasized the unprecedented nature of the fire. Again and again, observers lamented the fact that, while coal mines exploded frequently, quartz mines had always been thought comparatively safe. The disassociation between quartz mines and fires was so strong that one newspaper even mistook Hollinger for a coal mine in its report: "47 men are believed to be dead through a fire in the Hollinger coal mine, Ontario, Canada," reported New South Wales's *Barrier Miner*. The story of the Hollinger fire unsettled ideas about the comparative risk of gold mine work. As one Tasmanian paper put it, "A fire in a quartz mine [was] previously unknown."[68] Before Hollinger, the idea that gold mines could catch fire would not have readily occurred to stakeholders in any of the world's major mining zones. The world's mining states now turned a wary eye to the ticking time bombs potentially awaiting workers under their own soil.

Meanwhile, Godson's lengthier and less accusatory report on the lessons from the Hollinger fire proved most palatable for regulatory governments and corporate management. For example, an American Bureau of Mines circular by Daniel Harrington on mine ventilation used the Hollinger fire to call attention to the fact that metal mines around the world "had been of the opinion that [metal] mines were essentially immune to fire." The practice of dumping garbage in stopes was "a practice which has been more or less prevalent (and one of whose probable dangers would be, in fact have been, on numerous occasions, scornfully minimized) in metal mines not only in Canada but in various parts of the United States." In addition to the Hollinger example, the circular cited a 1928 metal-mine fire in Mexico that killed twenty men (but remarkably destroyed only about three cords of mine timber) and discussed the problem of open lights (for smoking or for lamps) and unsafe dynamite handling. Harrington then reproduced Godson's entire list of recommendations for his American audience as part of his call for increased state regulations of ventilation in metal mines and additional corporate attention to ventilation problems.[69] His aggressive campaign for mine fire safety resulted in a number of similar reports through the late 1920s and into the 1930s. An article in the *Canadian Mining Journal* in December 1928 called on mine managers to make basic adjustments to their operations to improve safety, including enforcing no-smoking rules and safety doors. The piece referred to Harrington's extensive American experience and the international conversation sparked by the Hollinger fire: "There have been a number of misconceptions in mining circles concerning the affair," Harrington wrote, before adding his own opinion on the matter: "It was caused in my opinion by an open light," possibly "a match from a smoker," but more likely "due to a very reprehensible habit which persists in metal mines that in order to obtain information as to what may be the condition in a winze or a stope which is below. . .the metal miner frequently lights a piece of paper and throws it into the opening."[70] The article implored mine managers to learn from the history of metal-mine fires in the United States and Canada, exert control over the dangerous habits of their employees, and modernize their infrastructure to save lives.

Data collected by Harrington on the Hollinger fire (among other reports on mine fires collected through the 1920s and 1930s) informed the official recommendations of the Safety Division of the US Bureau of Mines. The Safety Division conducted education and training for nearly ten thousand workers and mine officials, collected safety reports from engineers and mine operators, disseminated data on mining safety, visited fire sites, provided resources for administers and government officials, and ran International First Aid and Mine Rescue Meets across the country.[71] Although the onset of the Depression slowed widespread adoption of legislation on metal-mine fires in the United States, some adopted new laws based on the Safety Division's recommendations by 1930.[72] Harrington's extensive reporting on fires in metal and coal mines would later go on to shape American legislative change in the United States in the 1940s.

Waste and Environmental Consciousness

In the context of waste accumulation above- and underground, a nascent environmental sentiment found purchase at Porcupine. In the early years, conservation was sold to Porcupine residents in the context of forestry on the premise that it could help prevent forest fires—a powerful argument given the town's history.[73] Later, Porcupine locals engaged active outdoor lives in line with emerging values predominant in the United States and the urban South. A road north of Cobalt through the Temagami Forest Reserve had been built for tourist traffic coming north in newly popular automobiles (with the mines as an attraction).[74] An emerging cottage industry at Nighthawk Lake and a Bayside Beach Resort with "dancing, boating, bathing, fishing, tourists' huts for Rent, Camp Sites, etc." sprang up at Frederick House Lake.[75] Residents participated in dog races and long-distance skiing in the winter, and mining companies sponsored baseball teams in the summer.[76] Sportsmen extolled the values of the movement's principles for preserving game in the area. One of these men published a long editorial in the *Porcupine Advance* asking that Frederick House Lake be set aside as a feeding and breeding place for wild ducks.[77] In 1929 the *Advance* printed an article about the efforts to preserve migratory birds in Canada that referenced the American conservation movement and argued that Canada take similar actions.[78] That year, an active rod and gun club worked with the Department of Game and Fisheries to stock local waterways, including Pearl Lake and the Mattagami River.[79]

Given the thoroughness of gold mining's integration into the landscape and the way its wastes were spreading by the 1920s, conflict with these emerging conservationist impulses could not be entirely avoided. This was especially true in Porcupine's contested waterways. In 1923 fishermen noticed that the new dams on the Abitibi blocked spawning sturgeon from traveling up the river. In response, the Hollinger built fish ladders that would allow the animals to pass upstream from their power plants. In the aftermath of the case of *Hollinger v. Northern Canada Power,* fishermen requested similar modifications on the Mattagami. A decade later, conflict between the mines and the city over drinking water became an electoral issue. In order to supply enough water for everyone, new water mains, a new pumping station at Gull Lake, and new steel pipes brought water from McTavish Lake to Gull. At no point in any of these conversations did mining's presence on the land come under critique. Rather, amicable solutions satisfactory to conservationists and companies alike dominated. Where the combined pressure on resources became too much, technology provided solutions for managing resources in a way that, on the surface, satisfied everyone.

Yet in the places where amicable technological solutions were impossible, the mining industry always took precedence. In the midst of the conflict between Dome and Digby Vet over tailings disposal, Henry Depencier received a curt letter from the council of the township of Tisdale inquiring about rumors that Dome planned to pump its tailings into Porcupine Lake. Depencier replied, untruthfully, that "it is not the present intention of this company to pump our tailings into Porcupine Lake" and asked the council to "inform me as to who intimated. . .that it was our intention" to do so. Of course, Dome *was* planning to use Porcupine

Lake for waste disposal and was prevented from doing so only by the overlapping industrial uses of the lake that made rights too expensive to obtain. In 1924 the question of use of Porcupine Lake rose for a second time. Perhaps because it was one of the few local water bodies still uncontaminated by waste, mine employees had begun using it for a variety of domestic purposes. Now, the township wanted to use it as an outflow for South Porcupine sewage. In a letter to his lawyer Alex Fasken, Depencier observed that "Porcupine Lake water is used throughout our [mining] property for washing, laundering, etc." South Porcupine's sewage plans threatened current usage, and Depencier wondered "whether we have any rights at Porcupine Lake in this connection." If the sewage plans proceeded, Dome would have to implement "some system of purification in order to be safe, and such an operation may be quite complicated at our plant."[80] His company's needs prevailed. In another example in 1929, a settler sued Hollinger for damages after the mine polluted a stream on his property. The case appealed to notions of water rights that had forced the retraction of hydraulic mining in California fifty years before. The suit failed.[81] Yet at no point was environmental protection or outdoor recreation seen as antithetical to extraction in Porcupine. Rather, alternative uses for the land would be accommodated where possible and counted as wins for both the mines and the community. Where alternative uses proved incompatible, mining quietly dominated by wielding economic, legal, and rhetorical power built over the past two decades.

Waste and Its Consequences in the 1920s

As waste accumulated above and below the ground with serious consequences for communities and workers, it became increasingly difficult to ignore. The expanding scale of industrial mining widened the gap between the mines and the communities partly because the consequences of extraction fell unevenly. Workers and communities used the tools at their disposal to better their working and living conditions, but they always did so within the wider interests of the extractive industry. Because workers' bodies were the medium through which mining companies achieved their purpose, they could be blamed for systematic failures in the mining system. The rhetoric around the Hollinger explosion absolved the company of guilt and shifted the focus of the discussion to workers and their actions. Meanwhile, the industry framed its failure to deal with aboveground waste as inevitable and inherently unsolvable. Depencier's utter uninterest in where his company's wastes ended up suggests a context in which mining had no obligation to answer to its impacts (until it impacted the business of other mining companies). In all cases, the question of whether mining companies should be allowed to deposit waste on the land at all never entered the conversation in a substantive way. This idea was reinforced by the relationship between the mining companies and the nascent conservation movement at Porcupine, whereby the mining companies always dominated where technological or scientific solutions satisfying both sides could not be found.

As they confronted the waste heaps accumulating in their empty stopes and at the unprofitable ends of their mills, Porcupine's corporate elite extended a pattern that discard scholars and environmental historians have already noticed in their studies of other industries around the world. As Joel Tarr writes of urban environmental pollution, the "technological fixes for one environmental problem have often produced difficulties in other domains."[82] In many ways, the story of waste management is the story of the constant search for adequate receptacles (Tarr calls them "sinks") for the unwanted by-products of extraction. Yet tailings ponds and trash-filled stopes were also the site of *spills,* which, as Jennifer Garbys argues, are a symptom of the embeddedness of systems within their environments.[83] In the Abitibi, mining systems designed to benefit investors created waste that overflowed onto the physical environment and changed its function in unpredictable ways. It is little wonder that environmental historians trace the earliest environmental conflicts in the United States and Australia to mining waste in the aftermath of the California and Victoria gold rushes.[84] Mine pollution also drove international conflicts over watershed pollution in places like the Columbia River valley and the Elk River.[85] Tailings disposal would continue to plague mines and their regulators, and incidents like the 2014 Mount Polley disaster show that the problem of mining waste remains pertinent more than a century later.[86] Then as now, technological solutions for mining waste focus on damage control and ensuring the continuity of the industry's ability to dispose at the lowest possible loss to investors.[87] When understood in the context of the Mattagami reservoir land appropriations, the accumulation of wastes above- and below-ground as an extension of colonialism in northern Ontario. Although they did not appear in settler discussions around waste management in the 1920s, the pollution of the land infringed on Indigenous sovereignty in the Abitibi in ways that continue to be felt in the present.[88]

As Porcupine grappled with shared problems of industrialization, international audiences started looking to Porcupine for answers to tricky new industrial questions. They often borrowed people, technology, and, in the case of the mine rescue cars, physical objects from their international counterparts. This flow was not one-way. Canadian miners, managers, and officials found themselves at the center of new conversations about mining and environmental health—above and below the ground. In general, northern Ontario's mines preferred technological solutions that did not threaten the continuation of industrial extraction, preserving the hegemony of the mines and their economic order. Although we can see suggestions of this trend in the history of tailings, mine safety, and conservation at Porcupine, the industry's approach to environmental problems would take on new power and clarity in the context of industrial lung disease and the silicosis crisis.

CHAPTER FIVE

World of Dust

The Rise of Canadian Silicosis Science

On October 6, 1924, Dr. A. R. Riddell conducted an autopsy on Canada's first official silicosis victim. Two months earlier, Riddell had diagnosed his now-dead patient as silicotic. He had based his diagnosis on X-rays that had shown dark cotton-like shadows in the man's lungs and an interview (conducted between coughing fits) telling of a history of work in a dusty crusher room. Now Riddell would see firsthand what prolonged exposure to Ontario rock dust could do to a human body. As he cut the chest open, swollen lungs bulged out. A dense fibrous tissue coated their surface, one and a half centimeters thick in places. The upper part of the organ was bluish gray in color. It should have been spongy and light, but it was so packed with dust that it sank like a rock when Riddell submerged it in a beaker of water. Glands had been nearly obliterated by fibrosis and were so hardened that they were difficult to cut. Lab analysis of the silica content in the dead man's lungs showed they contained five or six times what Riddell had even expected.[1]

Understanding the environmental history of mining in the 1920s means recognizing that the stakes of overcoming crises had become extremely high. In the pursuit of ever-lower grades, the mines had become industrial behemoths. In 1928 Hollinger employed 2,500 men, 1,540 working underground. The mine included 100 miles of underground drift, 8 miles of shafts, consumed 835,000 gallons of water per day, and produced 370,000 tons of waste rock per year. Meanwhile, work commenced on 156 working stopes, while 113 stopes waited to be filled, 29 stopes were in the process of being filled, 153 stopes partially filled, and 42 stopes filled. All this to produce 6,000 tons per day of ore. Powered by ever-rising numbers of men, money, and machines, the mine rumbled on through time and space with an incontrovertible momentum. Mining was the North's purpose. Unexpected environmental problems threatened to bring down everyone with personal, financial, and ideological stakes in their production. In this context, bureaucrats, scientists, and corporate directors were highly motivated to find quick solutions to keep the gold coming out of the ground.

To this point, fire, water, and geology have dominated Porcupine's environmental history, but concern about dust had long lurked under the surface. To its credit, the Ontario Bureau of Mines had pushed for wet drilling and proper ventilation starting at its inception in 1891, although their sermons must have seemed absurd to the tiny developing mining companies scraping moss from rock with pickaxes under open air. In South Africa and the United States, workers and their governments sounded alarm about new industrial diseases in the gold mines. Yet the idea that Canadians were immune from the problems of other global mining giants endured even as Canadian companies expanded in the early twentieth century. Despite their size, Canadian companies saw themselves as still in the development stage of sophisticated industrial extraction, struggling with preindustrial frontier problems of basic survival in a harsh and unpredictable climate. Compared to South Africa, where silicosis was prevalent, Canadian mines were young, the mining less deep, the rock of a different character, and the workforce small and healthy. Awareness of the danger at the regulatory level dawned very late. In 1924 the body on Riddell's autopsy table became the first indication of impending disaster.

Insidious and nearly invisible, silicosis became a slow-moving but intractable problem in Canada.[2] Brave northern men could not fight silicosis as they had the Hollinger fire, and clear lessons for future prevention proved difficult to discern. Unlike the other environmental crises mining had faced, silicosis could not be solved (or even conveniently obscured) by the application of predeveloped international technology. By the time the disease arrived on the Canadian soil, other mining states had already spent two decades trying various iterations of regulatory, scientific, and social control without much effect. Instead, conditions had actually gotten steadily worse, eating into corporate profits, undermining labor relations, and deepening chronic labor shortages. Meanwhile, the rise of industrial hygiene in the early twentieth century increasingly framed industrial risk as unacceptable in the context of workers' rights.[3]

With no simple answers and much to lose, the Canadians turned their efforts inward to develop a scientific solution for their silicosis crisis. The product of their efforts was McIntyre Powder. McIntyre Powder was developed by a conglomerate of Canadian scientists, backed financially by the big mining companies of the Porcupine region, and then, after being tested on Canadian soil, sold as a silicosis miracle cure to mining communities around the world. Ultimately, however, McIntyre Powder functioned in much the same way as other technological "solutions" to environmental problems. With the application of McIntyre Powder, industrial extraction could continue without significant (and expensive) changes to social, political, and economic infrastructure. In fact, McIntyre just shifted the costs of silicosis away from the mining companies and onto workers.

The Emergence of International Silicosis Science

Like waste management, silicosis emerged directly from the advent of industrial mining. Gold is found in silica-based feldspar and quartz in the hard rock that makes up the Canadian Shield. Unlike other kinds of rock found in different types of mines, the rock in the Canadian Shield (and the South African Rand, Western Australia, the American Midwest, and other hard-rock zones) produces sharp and corrosive silica shards when miners break it down. When combined with the capital and labor structures of industrial mines, silica-bearing rock proved deadly. Within the confined spaces of deep industrial mine tunnels and working near dust-generating machinery like drills and crushers, miners breathed silica dust on every shift.

The scientific understanding of silicosis took time to coalesce. Before the turn of the century, silicosis had not been used to label miners' lung disorders. Instead, consumption, tuberculosis, or the "white plague" served as umbrella terms for any kind of chronic lung disease. The fact that miners were particularly susceptible to lung disease has been well understood since antiquity, and by the industrial era was referred to as miners' phthisis or miners' lung: broad categories that encompassed any kind of pulmonary distress related to life or work in dirty, damp, crowded, or dusty spaces. In 1882, Robert Koch discovered that some of these disorders could be attributed to the tuberculosis bacillus. Historians of silicosis argue that Koch's discovery actually delayed legislative and scientific action on silicosis because the bacteriological approach de-emphasized environmental factors. After Koch, physicians began associating any kind of respiratory distress with the tuberculosis bacilli. The scientific context aided the entrenchment of this association because tuberculosis seemed similar to other nineteenth-century bacterial/micrographical discoveries providing blanket explanations (and solutions) to common health problems. Physicians explained the connection between dust and disease by framing dirt as a vehicle for tuberculosis: Dust in the workplace or home might carry tuberculosis bacteria from person to person; in the mines, coughing and spitting on the ground resulted in the transference of tuberculosis between miners living and working in the same space.[4]

Not everyone subscribed to the bacteriological model. Although scientists would persist in connecting tuberculosis baccili with silicosis well into the twentieth century, the nuances had little impact on laymen and miners. The connection between lung trouble and dusty environments was obvious to the people actually breathing dust every day. Even within the research community, scientific categories generated in Europe did not always reach its colonies and successor states. Silicosis science thus remained uneven and, in some cases, contested between workers and experts alike.[5]

As time went on, tuberculosis proved increasingly insufficient as a catchall diagnosis. Lung diseases manifested in multiple environments among many different demographics and produced a wide variety of symptoms. To make sense of some of these complexities, British physicians began to differentiate more clearly between different types after 1900. One of the earliest mentions of "silicosis" appeared in the *British Medical Journal* in 1903. The author, Thomas Oliver, referred to silicosis as a variant of miner's phthisis found in quarrymen and gold miners. Oliver differentiated phthisis from tuberculosis by pointing out that the lungs of gold miners suffering from silicosis contained no trace of tuberculosis bacilli. Quoting work conducted by several English doctors, Oliver further argued that silicosis and tuberculosis were related but distinct. The silica shards seem to have provided entry for tuberculosis in some (if not all) cases. This was proved by comparison between coal miners and quarry workers from the same area. The quarry workers suffered from lung diseases, while the coal miners (although spitting up quite a lot of black phlegm) did not. Oliver hypothesized that the sharp silica dust in the quarry, unlike the softer coal dust, provided entry points for tuberculosis bacteria.[6]

Within this context of debate within the scientific community, industrial miners began dying on an unprecedented scale. South African politicians, mining companies, and scientists perceived the potential human and financial threat posed by silicosis first. Chronic labor shortages and the need to understand and control migrant workers provided favorable conditions for studying miners' health.[7] South African historian Jock

McCullough calls the period between 1902 and 1912 a "silicosis crisis" in the newly formed state. New dust-generating technology, increasingly confined underground spaces, and enormous workforces combined to create the perfect conditions for a silicosis epidemic.[8] The 1902–3 Weldon Commission identified silicosis as a major problem and recommended dust-reduction measures in the mines.[9] However, with no precedents to turn to and little political will, the commission accomplished little except increased surveillance of the medical condition of new labor arriving on the goldfield.

Nevertheless, by naming the problem, the commission set a slow process in motion. After 1914 mining companies subjected black miners to regular examinations, and the 1916 Miners' Phthisis Act stipulated examinations every three months. As with everything in the industrial mines, these examinations quickly got scaled up. The massive numbers of men examined under these programs required a kind of medical mass production. McCullough estimates South African doctors conducted a minimum of seven hundred thousand examinations per year and cites one doctor who bragged he could conduct a thousand tests in three hours. Exams relied on a miner's weight (any more than five pounds lost between examinations could be a basis for diagnosis) and X-rays (for whites only). These systems were not particularly effective for a variety of reasons, but especially because labor-hungry mines proved unwilling to exclude any potential worker who could pass a cursory exam.[10]

Standardized testing under the 1916 act established a stable biomedical understanding of silicosis. Pieced together using a combination of South African, British, and Australian research, the newly emerged epidemiology maintained the following: dust caused silicosis; it was difficult to diagnose (particularly in its early stages); there was a strong relationship between silicosis and tuberculosis; and continued exposure to dust after diagnosis is always fatal. It also introduced distinct stages, each associated with a different level of debilitation and compensation. "Primary-stage" silicosis implied that the man had not been totally incapacitated, whereas "stage two" silicosis meant the man was so sick he could no longer work at all. These categories were eventually expanded in 1919 to include "ante-primary" to describe silicotic men who could still work, but only if they left the mining industry.[11] As the first and only clear description of silicosis, this orthodoxy would come to be accepted widely across the world and formed the basis of compensation and employment policies almost everywhere men mined for deep gold.[12]

Australian research followed closely on South African foundations. Concerned officials and medical professionals pointed out high lung disease and death rates in Bendigo as early as 1900. Western Australia saw the first major inquiry into these alarming numbers in 1905, finding, in a report similar to the South African Weldon Commission, that the Kalgoorlie mines exposed miners to dust due to inadequate ventilation. Although the Western Australian government took measures to improve conditions, a 1910 report found that miners still suffered disproportionately from lung disease, which it divided diagnostically into early, intermediate, or advanced fibrosis or full-blown tuberculosis (or both). Lack of political will resulted in minimal reform before the 1920s, but Western Australia did institute an examination scheme similar to South Africa's as well as a modest compensation program for affected miners before the First World War.[13]

Officials in American jurisdictions followed the South African and Australian events, but proved extremely slow to adopt meaningful silicosis legislation. Unlike the Commonwealth countries, in which government commissions sounded the alarm, American insurance companies spoke up about silicosis first. In 1908 statistician for the Prudential Life Insurance Company Frederick L. Hoffman unmasked silicosis as a distinctly industrial disease threatening the sanitary and economic health of America. He was followed by Metropolitan Life Insurance Company statistician Louis Dublin, who publicized similar findings.[14] But within major medical journals and among doctors, the bacteriological explanation continued to dominate discussions in America, so these economic arguments enjoyed little official attention.

Eventually, European and South African studies combined with statistical materials about mortality in metal miners aroused enough concern to spark a US Public Health Service investigation (in collaboration with the Bureau of Mines) in 1911. The investigation resulted in the identification of silicosis as a major health hazard for metal miners. It was followed up in 1914 with a landmark investigation on the lead- and zinc-mining region of Missouri, Kansas, and Oklahoma conducted by Anthony J. Lanza. Lanza emphasized environmental causes for industrial disease and relied on worker testimony in addition to traditional laboratory tests. The Lanza method of collecting worker history became an important model for silicosis diagnosis (as well as other industrial diseases). He found high rates of tuberculosis and a close relationship between consumption and tuberculosis. Yet Lanza continued to emphasize the bacteriological factors in lung disease (especially living conditions) until further 1919 statistical work by Frederick L. Hoffman on miners in Barre, Vermont, finally demonstrated conclusively that lung disease persisted long after tuberculosis declined. After this point, silicosis became widely accepted as an industrial disease, forming the basis for strikes and expanded reform efforts into the 1920s.[15]

Silicosis research occurred in other locations besides South Africa, Australia, and the United States. Mexican workers' compensation, for example, engaged actively with the global silicosis debate (and its implication for miners) during the social revolution of 1910. New research from Latin American scholars sheds light on the different ways silicosis science was engaged in these contexts.[16] When Canadians developed their silicosis cure, they did so with reference to what was happening in Latin America and sought out places like Mexico and Peru as potential markets. This relationship is important to understand as part of the foundation of Canada's later imperial expansion into Latin American mining and its consequences for miners' bodies and local environments.[17]

Silicosis in Canada

As part of the international mining community, the Canadian industry participated actively in operational discussions around lung disease. Ontario's inspector of mines A. Sleight commented extensively on mine ventilation in the Ontario Bureau of Mines' first annual report in 1891 and exercised his authority to force some mines to improve air quality underground.[18] In 1912 Mining Commissioner S. Price's report on the benefits of an eight-hour day directly and prophetically linked dust to miners' phthisis. On the whole, Price disagreed with miners' assessment of underground work as unnatural and unhealthy: "The mines in Ontario, I believe, are naturally as healthful as any in the world." Nevertheless, he thought that there probably was reason for concern around the inhalation of dust from drilling. Although phthisis "is at present a disease little known in Ontario," it was only a matter of time and development before it started to become a problem. Besides, he noted, even if miners were not dropping dead from dust, it probably was not good for them. Reducing the number of hours exposed thus might improve miner health. He noted that inspectors knew about phthisis and were keeping an eye on the problem.[19] Indeed, in 1912 the Bureau of Mines cited improper ventilation of mines as one of seven major ways mining companies failed to abide by mining regulation in Ontario, contributing to increased mining accidents.[20] In a long write-up on the topic, Inspector E. T. Corkill wrote that "little attempt has been made in the mines of Ontario to adopt any form of artificial ventilation," even though "in recent years. . .owing to the extent to which some of the metalliferous mines have been worked," artificial ventilation was probably necessary. Corkill directly copied Transvaal ventilation standards into his report. He went on to cite evidence from the East Rand in South Africa and the Comstock mines in Nevada to conclude that "the installation and operation cost of any of these at the mines in Ontario. . .would be comparatively small, and would be insignificant compared with the valuable results obtained." Corkill implied here that increased worker health and safety equaled corporate profitability for the industry, but left the reason for instituting changes (worker health) unnamed. Labor publications, including the federal *Labor Gazette,* also monitored miners' health and specifically paid attention to instances of miners' phthisis.[21]

The Ontario Bureau of Mines first acknowledged the presence of miner's phthisis in Canadian mines in 1913. In his inspector's report, Corkill demonstrated a keen awareness of the disease's problems elsewhere in the world and admitted ignorance as to its relevance on the Canadian field. From his previous engagement he knew that the disease associated with quartz and that increased use of hammer drills would make dust "a serious menace" unless preventative measures were taken. In 1913 these measures meant simply ordering mines to equip drills with a water spray, which Corkill did. Ontario was the first and only Canadian jurisdiction to take such measures, and as a result Porcupine became the front line for Canadian dust legislation.

A year later, Corkill admitted that his attempts to force mines to use water sprays had largely failed in the face of considerable resistance from several fronts. The water-spray attachments on hammer drills frequently failed: "The spray chocked [*sic*] up easily and required considerable care to keep in good working order." Not only that, but miners had to carry water up the raises and stopes to get it to the drills, and often unintentionally got the driller wet. The pace of drillers determined the amount of ore processed in the mines, and water "undoubtedly decreases the footage drilled, [so] the foremen and managers probably gave more serious consideration to these objections than they otherwise would have done."[22] By 1915 spray attachments had been abandoned.[23]

The consequences of Ontario's failure to deal effectively with its dust problems had not yet become obvious, but Corkill knew it was only a matter of time. In these prewar years, "there are not yet many cases of miner's phthisis," he wrote, but it was difficult to tell what the effects had been because "when men become sick at the mines or feel unable to work, they leave at once for their old homes and are lost track of."[24] The problem of tracking the health of seasonal, temporary, and transient workers was something Ontario shared with South Africa. The Transvaal already bore the distinction of being a leader in phthisis management, and Corkill reproduced several pages of findings from the Mining Regulation Commission of the Transvaal under the "Health of Miners" section of his Ontario report.[25] The piece he chose argued unequivocally that the phthisis experienced by the miners was silicosis, not tuberculosis, but that it could be complicated by tuberculosis if the bacterial infection "becomes superimposed" on the damage caused by rock dust. The report cited statistical studies in Cornwall (England) and Bendigo (Australia). Corkill's choice to include this large section is indicative of his perception of Ontario as belonging to a wider collection of industry giants susceptible to the same problems. Perhaps his audience had not got the point, however, because Corkill reproduced the same report again in 1914, this time adding details from a silicosis publication from the *Rand Daily Mail*.[26]

Corkill's warning remained mostly unheeded outside of the bureau. When the Workmen's Compensation Act came into force on January 1, 1915, it deeply disappointed him: its single provision for silicosis seemed a poorly thought-out side note. The act stipulated that the mines paid an annual amount based on the board's assessment of accident compensation costs, and this amount was then used to pay out sick or injured workers (or paid out to their families in the case of death). Industrial disease (it specifically named miners' phthisis) was considered a personal injury by accident. Corkill saw this as a singularly ineffectual way to address the disease. Phthisis did not behave like other accidents. It was chronic and degenerative and could take years to appear. Employers were required to notify the board by mail within three days of an accident, but in the case of phthisis there was no "moment" at which it occurred, and a man might not even know he had the disease until many years after leaving the industry.[27] The new act focused on obvious and instant accidents and was totally unequipped for dealing with the slow degenerative nature of miners' phthisis. Corkill also complained that the act did nothing to prevent foreign miners already carrying the disease from entering Canada and taking advantage of compensation, although he had no numbers to back up his suspicions.[28] Meanwhile, he continued to follow the progress of South African research. In 1915 he cited findings from the interim report of the Miners' Phthisis Committee of the Union of South Africa, which argued that water was the only way to prevent dust and also emphasized the dual threat of silicosis combined with tuberculosis.[29] He could see the applicability of this research for Ontario, and the new compensation laws seemed an unsatisfactory answer to what had clearly become a major problem on other goldfields receiving much more attention than it did in Canada.

Thus, the alarm bells sounded by South Africa, Britain, Australia, and the United States resulted in little tangible action in Canada before the 1920s. The cumulative and degenerative nature of dust exposure meant that an incubation period of a decade or two of sustained industrial mining needed to pass before the disease's effects became visible in the mining population. South Africa and Australia had already witnessed several decades of underground hard-rock mining before they recognized silicosis as a problem. In contrast, hard-rock gold mining did not begin in northern Ontario until 1909. Compared to regions already beginning to worry about silicosis at the turn of the century, Porcupine was a new field. Then, for the first few decades of extraction, environmental and human factors kept the mines comparatively small. High turnover of men and mining companies, fires, floods, wars, remoteness, and other problems kept men aboveground and out of immediate danger. Even as extraction expanded, this context of constant crisis management left little room for poorly understood problems like silicosis, especially since no one could see or measure its effects in these early years.

Besides, there was the matter of accepted risk in the mining industry. Historians of mining in the United States note that, in the early twentieth century, environmental danger and lung disease constituted normal (and perhaps even acceptable) risks associated with the work of mining. These conditions would become problematized only after the war with rising awareness and concern about industrial disease and worker rights.[30] Even then, workers could get caught between those who saw silicosis as an industrial disease and those who attributed it to poor sanitation and moral depredation—matters of individual responsibility. Miners' relative invisibility (underground, in remote locations) compared to other industrial workers made the debate even more amorphous.[31] In Canada the sense that the mines remained a risky place despite the best efforts of government, operators, and miners themselves remained prevalent throughout the 1920s. Even at this late date, popular opinion was that the best anyone could do was "provide for constant watchfulness and care."[32] Furthermore, even into the 1930s there was a moralistic tone to silicosis diagnoses. In a letter to Depencier in 1933, the secretary of the Ontario Mining Association, G. C. Bateman, wrote that Dr. Lanza of the American lead- and zinc-mine investigations had confirmed that "syphilis may be an important factor in the development of silicosis" and recommended that blood tests become part of the routine examination conducted on Ontario miners.[33] By linking illness to the miners' moral and physical depravity, Bateman rhetorically shifted blame off the shoulders of his company and onto the miners themselves. If silicosis was a question of lifestyle, it was unclear who was responsible for compensating its victims: individuals, the company, or the state.

The transient nature of work in Ontario mines also slowed recognition of the disease. As "unskilled" generalists in the industrial mine model, miners moved around a lot—not only from job to job, but from mine to mine and from occupation to occupation within a mine. Few recorded silicosis victims had been lifetime miners. The biographies taken by doctors chronicle this occupational mobility: For example, an Italian who arrived in Canada in 1906 worked on the railroad for six years. When the mining boom began, he found employment at Porcupine. In 1912 he worked doing odd jobs on the surface, then began as a mucker for one year, ran a hoist for six months, was a machine helper for one year, before finally moving into machine drilling, where he contracted silicosis. Another, a thirty-five-year-old French Canadian, worked at "bush work" from the age of thirteen, then for the logging industry as a log driver until he was twenty, before turning to prospecting, which he did for six years. When he arrived in Porcupine, he engaged in mixed mining and prospecting before finally settling in as a timberman and eventually an underground shift boss.[34] This kind of mobility slowed the progression of the disease and made keeping track of victims difficult.

So it was that Ontario finally entered its silicosis crisis with the Italian miner on Riddell's autopsy table in 1924. Increasing attention to workers' rights and escalating postwar production combined to highlight instances of lung disease among Ontario miners. As the case numbers rose, the regulations in place under the Worker's Compensation Act proved inadequate. The international context indicated that the potential solutions would be complicated and expensive because silicosis was an economic and environmental issue not easily detangled from the structural and ideological underpinnings of industrialization as a whole.

Canadian scientists and the government reacted first, the mining companies only joining in the general panic after 1926. Riddell was not the only one working in silicosis in 1924. That year, Jabez Elliot wrote a paper for the *Canadian Medical Association Journal* that chronicled the research in South Africa, Australia, Missouri, and Vermont and noted that Ontario mines, expanding rapidly, had been digging into geologically similar silica ore. Ontario schist did not shatter like quartz. In the past, visual assessment of the rock's smoother edges led the industry to believe that their rock did not cause the same kind of damage inside miners' lungs. Examination under a microscope showed, however, that Ontario schist rock still had "uniformly jagged and sharp edges and are occasionally needle-like in form."[35] On the request of the Division of Hygiene in the Provincial Health Department, Elliot gathered slides from X-rays of miners' lungs. Dr. R. R. Sayers, chief surgeon of the US Bureau of Mines, and Dr. Henry Pancoast, a consulting physiologist from the US Bureau of Mines, examined the slides, reflecting the international nature of Canadian silicosis work.

Elliot was careful to eliminate any miners who had worked at Porcupine for less than five years, mined at other camps, or did not work underground. These stipulations were meant to show definitively that miners could contract silicosis from Ontario rock, rather than having contracted it elsewhere. Elliot eliminated all but eleven miners based on these criteria.[36] In the tradition of the groundbreaking silicosis studies in the United States and South Africa, he combined personal history taking with X-ray and physiological examination.

The results of the study were alarming. Only two of the miners showed total freedom from silicosis. Of the remaining nine miners, four showed clear evidence of silicosis; three showed fibrosis, "which may or may not have been due to dust"; and two showed evidence of early-stage silicosis.[37] The ethnic divisions inherited at Porcupine since the war loomed large in this small sample. Of the four confirmed cases, two were Italian, one Finnish, and one French Canadian. All had varied experiences underground mucking, timbering, driving locomotives, drilling, or scaling. The French Canadian had been an underground shift boss, but the other victims all worked difficult and dangerous underground mine jobs. The occupational segregation of work meant that some groups would bear the burners of silicosis most acutely.

Elliot had also examined the physical presence of dust in the mines. The methodology was strongly influenced by American studies, to facilitate direct comparison to American mines. Meanwhile, South African standards provided an acceptable base line: 300 particles per one cc of air, or 7,680,000 per cubic foot. Compared to their international colleagues, the Canadian mines came out rather well. The only place where dust was above "acceptable" levels (defined by international standards) was in raises, where wet drilling proved insufficient for keeping down dust. Although Elliot argued that the mines seemed to be doing well in terms of avoiding dust so far, he warned that the problem would grow worse as the industry developed. In his conclusion he recommended annual examinations for all men employed underground for more than five years, increased ventilation in the mines, and the discontinuation of all dry drilling. The mining industry was listening, although it would not realize the extent of the crisis for two more years. In 1924 the Dome Mining Company bought its own X-ray machine.[38] This was housed in the compressor room until 1929. Experts from Toronto (Dr. Haig and Dr. Bain) used it to examine miners. The X-ray had a long life in the mines and was used to test thousands of miners in Porcupine over the next several decades.[39]

In the summer of 1925, the Ontario Department of Mines sent its chief inspector, T. F. Sutherland, to South Africa to study mining conditions. He was specifically instructed "to report upon. . .accident prevention and silicosis." The choice reflected South Africa's perceived status as a leader in terms of silicosis research and treatment. Broadly, his trip was promoted as part of the government's effort to deal with the increasing size and depth of the Ontario mines. "We have not yet had. . .much deep-mining in Ontario," Sutherland told the *Globe* in the days leading up to his trip, but "both the Hollinger and McIntyre gold mines are pushing down their shafts, the latter having a present objective of 4,000 feet." The hazard involved in deep mining "increases with depth: cables require to be heavier and stronger, hoisting machinery more powerful, and the difficulties of ventilation increase." Heat and pressure underground represented additional dangers, and although "Ontario's present regulations are based on South African experience. . .they are long standing, and in many cases new methods have come into use."[40]

In his report, Sutherland compared Ontario to South Africa. South

African deposits tended to occur horizontally (whereas Ontario's were vertical), so "gravity accidents" tended to be less frequent on the Rand than in Ontario. Yet Sutherland believed that "the fact that nine-tenths of their labor is Kaffir [*sic*]" made the underground conditions more dangerous in Africa than Canada—a racial argument echoed (on a much smaller scale) in Ontario around safety and non-English speakers.

He also noted that prosecution for violations to workplace safety measures in Africa was much more effective due largely to the fact that specific men were made responsible for workmen. "In Ontario," he wrote (with the weariness of an experienced mine inspector), ". . . if a stope is not properly scaled, the mine foreman says the shift boss is responsible; the shift boss claims the scalers are responsible for that work, and the scalers say they had not time to get around to it and anyway the machine men are supposed to scale before setting up. Against whom can you take action?"[41] Responsibility in Ontario was not clearly defined. The result was that in South Africa, there were a total of 1,673 successful prosecutions in a labor force of 208,000 (1:124), whereas in Ontario there were 2 prosecutions in a workforce of 12,500 men (1:6,250).[42] The difference reflects Ontario's much later large-scale industrial development. For many years, Ontario's mines remained small. The safety of a few miners and prospectors was much more easily monitored and managed. In the newer industrial mines, safety became a complex operational task. As mining's scale ballooned, safety became a collective problem requiring top-down intervention, especially when the financial interests of the company (and therefore the pressure it put on the miners) prioritized speed and volume.

But the most important reason Sutherland had gone to South Africa was because of silicosis. In his luggage on the way home was a copy of South Africa's Silicosis Act, subsequently filed at Queen's Park for reference. Sutherland reproduced large sections of it in his report, with particular emphasis on examination schedules and compensation schemes. He also brought home a copy of a 1924 pamphlet by the Transvaal Chamber of Mines outlining the research done to date on phthisis in South Africa. The pamphlet emphasized water, ventilation, and dust sampling as the most important preventative measures. Finally, Sutherland brought back dozens of blueprints and reported on the practices currently in favor among South African mining engineers.[43] These were not limited to observations about silicosis. This part of the report wandered away from silicosis and even safety generally to explore everything from rope design to compressors to furnaces to welding techniques currently in vogue in the Rand and suggest Sutherland's (and the Ontario government's) fascination with South Africa's cutting-edge practices of industrial extraction.

Meanwhile, silicosis and silicosis research continued to progress on Canadian soil, with northern Ontario on the front lines. In the latter part of 1925 and the early months of 1926, the Division of Industrial Hygiene and the Provincial Board of Health conducted a survey of workers in Ontario mines and concluded, to no one's surprise, that silicosis was a problem. In 1926 the Ontario Bureau of Mines announced that "in view of the seriousness of the [silicosis] situation," it would begin remedial and preventative measures following South African practices.[44] The work of Riddell, Elliot, Sutherland, and others contributed to this effort.

The 1926 accident report for the Ontario Bureau of Mines reproduced Sutherland's 1925 report along with a series of research papers from South African and Canadian experts.[45] No specific recommendations or implications for Ontario were identified for special consideration. It seemed the bureau intended to copy wholesale the science and recommendations of South African Research. Indeed, on April 8, 1926, the Workmen's Compensation Act was amended to include more specific language for silicosis that separated it from other types of compensation. The language exactly copied the definitions of the silicosis stages as they were articulated by the South Africans. By that year, the board had accepted fifty-three cases in the ante-primary stage, forty-one in the primary, fourteen in the secondary, and fourteen with silicosis complicated by tuberculosis. The year before, it had compensated its first case of death by silicosis.[46] This new, separate system of compensation reflected silicosis's special status within the category of industrial disease, as well as the perception among legislators and workers' compensation advocates that the disease was somehow different from other ailments.

With these changes to the Compensation Act, silicosis finally hit mine operators in their pocketbooks. Suddenly, the disease became much more widely discussed. The community had seen it coming: people had seen (and even been tested by) the Toronto doctors traveling in and out of Porcupine to take chest X-rays and life histories. Although rumors flew, few understood the disease (or its implications) very fully. Dr. Hague held a public address at the Kiwanis Club in Porcupine in July 1926 to address some of their concerns. Here Hague explained in layman's terms how dust might be taken in to the lungs of men working in enclosed spaces. He touched briefly on the tuberculosis bacilli, its relationship to silicosis, and (demonstrating considerable awareness of the progression of lung-disease research) its role in impeding understandings of silicosis. He also acknowledged the recent presence of foreign doctors and other experts at Porcupine and the alarm their presence sparked in the community. He told the audience not to worry, because "many examined here had been found entirely OK." He finished his talk by showing slides of diseased lungs, including one of a tuberculosis victim whose body had begun to heal itself.[47] The overwhelming sense was one of reassurance and control.

Hague's talk emphasized the cooperation between the government and the mining industry on the silicosis matter, but the truth was that the government led Ontario's response. Silicosis's seriousness had only recently dawned on the mining companies. In 1926 they took their first step toward action. On November 3 and 4, the Ontario Mining Association pulled together a Silicosis Committee, which met for the first time in Hollinger's offices in Porcupine. Representatives from companies across the gold zone were present, along with members of the Workmen's Compensation Board (V. A. Sinclair, chairman; Dr. Bell, chief medical officer; and Mr. Graham, head of claims) and two men from the American Bureau of Mines in Washington (Dr. R. R. Sayers and Mr. D. Harrington). Two other doctors with silicosis experience (Haig and Bain) also sat in.

Until 1926, silicosis compensation had been relatively haphazard. Sinclair pointed out that some men who had received a payout for silicosis continued to work underground in dusty conditions, even though the payout was intended as compensation for lost wages following disability, and some even continued to receive awards as their conditions worsened. Not only that, but there was no agreed-upon method for determining whether a man suffered definitively from silicosis and no way for a mine to know if a man they hired had already been diagnosed with the disease. Furthermore, silicosis compensation came out of general funds. This was a problem because silicosis only affected miners in the gold industry, and therefore a special assessment above and beyond what the mines already paid for accident compensation was recommended.[48]

If the compensation process thus far had been ad hoc, the Workmen's Compensation Board was nevertheless more motivated to assist men with silicosis than mine management. The minutes from the Silicosis Committee's meeting suggests that the companies were caught flat-footed. Like tailings waste and dynamite boxes, silicosis had not been something they had thought much about. They had no statistics of their own, knew very little about the disease and its progression, and relied almost entirely on the board representatives and the doctors in the room. On the advice of Sinclair, following the 1926 meeting, the companies hired two doctors (Haig and Bain) recommended by Dr. Bell of the Public Health Department for $1,200 per year. They would referee silicosis cases because the X-ray machines alone, according to Dr. Sayers from the American Bureau of Mines, were unreliable. The films must be taken in conjunction with the history of each man and his individual physical condition. This, combined with ventilation and water, was the only way to manage the situation. Sayers also noted that the Porcupine X-ray films closely resembled those he had seen from South Africa and had fewer similarities with the Broken Hill slides collected in Australia. The committee expressed interest in obtaining the services of "Dr. Irvine" or "Dr. Smith," famous South African doctors experienced in silicosis diagnosis.

Based on the assessment of silicosis cases thus far, the secretary of the committee presented a silicosis index that calculated fair rates of assessment based on incidence of silicosis detected in the different mining camps. Rate amounts were calculated based on the percentage of silicosis cases found in each area. Although the sample sizes were small and irregular, the committee needed a quick financial solution to its looming compensation problem. Members agreed to the board's assessment—on the understanding that this was a temporary arrangement subject to revision following better silicosis research. From this point on, the mines began keeping careful silicosis records. The Dome Fonds, for example, contain multiple boxes of silicosis records starting after 1926. By 1930 the firm had its own statistics to employ in discussions on compensation and regulation of the disease.

As the mines struggled to come to terms with the invisible threat within their midst, social inequalities established in the previous decade structured parts of the debate. In 1927 a memorandum on Dr. Haig's examination technique noted that at one point two Finns were examined. Only one of the men spoke passable English, which was a major problem considering the central role of occupational history in diagnosing silicosis. Haig relied on translation, and as the memorandum noted, "some mistakes might be very apt to occur" in such circumstances.[49] In (undated post-1929) meeting notes, Dr. N. H. Russell, medical officer in the Porcupine District for the Workmen's Compensation Board, was "questioned as to whether or not there was any relation between nationality and incidence of silicosis and tuberculosis." In response, Russell "stated that Finns and Italians were apparently more susceptible."[50] He also recommended that Finns and Italians be subjected to more frequent and intensive examination than other miners.

South African expert Dr. J. M. Smith eventually did come to Ontario to research silicosis. He spent several months in northern Ontario examining miners and administrative systems and left, in 1928, with the somewhat relieving news that "conditions were not as bad as had been feared," but that all workmen should be examined and that a regular system monitoring the disease needed to be maintained.[51] This finding was lauded by the industry: the *Northern Miner* ran an article called "Silicosis Not So Bad as Feared" in the summer of 1927, following the annual meeting of the Mining Association that pointed out that the disease affected the gold mines, but not the silver or nickel mines and, further, that Ontario was not suffering nearly as badly as the Rand.[52] A year later, the journal published nearly the same article ("Silicosis Not Bad as Thought"), which argued that inexperienced Canadian doctors had overestimated the prevalence of the disease and had since been corrected by American and South African experts.[53]

But statistics told a different story, and deaths and new cases crept upward into 1930.[54] By the time Ontario's Dr. J. G. Cunningham returned from the League of Nations' first annual Silicosis Conference in Johannesburg in 1930, eighteen more silicosis victims had died and hundreds more identified as suffering from various stages of the disease. By this time, the mines' silicosis policy almost perfectly mirrored South Africa's: Ontario exactly adopted the scheme of diagnosis and the recommendations of the Miners' Phthisis Bureau of South Africa over the course of its silicosis saga.[55] Government fonds abound with communication between government offices on the subject of silicosis and compensation, especially between Canada, the United States, and South Africa.[56] Ontario sent copies of its silicosis legislation to foreign governments for reference and collected literature and statistics from abroad.[57]

Silicosis was a thoroughly international event. Its wide impacts allowed the government and the mining companies to collaborate on solutions, but it also created unexpected problems. In a series of correspondence between the provincial and federal governments in 1930, officials discussed the problem of miners arriving in Canada already suffering from silicosis. A Timmins-based correspondent for the Ministry of Mines wrote in 1930 that "there are a considerable number of miners coming in to this camp from the old country, who, upon examination, have been turned down for silicosis." Apparently, many of these men had worked in the South African gold mines, been compensated there, and then moved to Canada to find work.[58] The lack of clarity around the causes of silicosis played a significant role in enabling Canada to prevent the immigration of silicosis victims. Since silicosis was not in itself infectious, it was not clear grounds for denying entry to the country. In a letter to London, the Canadian government cited a research paper showing the association between silicosis and tuberculosis to argue that silicosis victims should be prohibited from entering the country: "Some pretty good authorities suggest that an infective element must always be present in addition to the dust to enable it to exert its fibrotic influence," reads a letter from the Canadian Immigration Medical Services. In deciding what to do about the issue, the first order of business was to find out how other countries were handling the situation: "It is thought that you would do well to begin your work by first getting information as to what the other dominions and the United States are presently doing, if anything, to prevent silicotic people from migrating."[59] British doctors followed this advice and heard back from Australia, South Africa, and the United States.

Indeed, if the government and mine executives tried to give the impression of control, this was a regime of crisis management. A post-1929 Dome memo referred to South African prices so high that "all future profits would be required to pay their silicosis liability." A future without profit proved threatening enough to spur action. Besides, the South Africans had advantages that the Canadians lacked, namely, a cheap black labor force that took the brunt of silicosis exposure at a low cost. "One fact we must not lose sight of," the memo reads, "the figures given for the Rand did not include amounts expended on native labor. . . . [W]ith us we have no such cheap native labor to be compensated for at lower rates so that every silicotic case we have means expenditure on the maximum basis."[60] Given the hefty potential consequences, a joint research project on Ontario silicosis based on the major companies' collective interest suddenly seemed prudent. The matter could not be left to the regulators to sort out on their own, especially when so many questions remained. How could silicosis be separated from tuberculosis victims? Should tuberculosis be considered an industrial disease? How could the companies keep silicotic workers from contracting silicosis and worsening their case (and thereby increasing the amount of compensation they were eligible for)? How could remote mines, which must depend on whatever labor came to them, regulate silicosis? Where would the money come from for compensation, sanatoriums, and education? The specter of massive compensation costs haunted the mines well into the 1930s and proved a powerful motivator for corporate action.

Aluminum Therapy and the Exportation of a Canadian Cure

Silicosis produced the first shreds of doubt about the ability of the international medical community to provide good answers for Canadian environmental-industrial health problems. The Silicosis Committee agreed on the fact that the results of all the cumulative research from the past two decades of silicosis in other nations "were not very definite." There was a clear sense that there was no blanket solution Canada could adopt from this body of international work. "South Africa was not wholly comparable to Ontario and it was felt that the Mining Association should make a study of the causes of silicosis and try and find a remedy. Medical men should consult geologists and mineralogists; they should examine slides to determine the shape and nature of dangerous dust particles and should not believe that South Africa and the United States have the only results that can be obtained."[61]

Ontario, it was felt, might be different. And at any rate, no one else seemed to have a good handle on silicosis. Finally, "The situation is serious enough for us to spend a substantial amount of money to try and determine the cause and remedy." At that point, Depencier from Dome pulled out information from Mysore mines in India, where they drilled dry, which showed that "the miners did not develop silicosis."[62] There were a lot of questions about silicosis that still needed answering, and the Canadian Mining Association felt that perhaps it was time to come up with its own answers. This moment was a marked significant change in the history of the industry, which had almost always exhibited the deepest trust in the knowledge, technologies, and people from its overseas community.

The silicosis story has a heavily ironic ending. Cases continued to escalate throughout the 1930s until the aluminum dust solution "McIntyre Powder" emerged from work conducted by the McIntyre Foundation (the research body that grew out of the Canadian Mining Association's desire for Canadian studies on silicosis). The McIntyre

Foundation research showed that aluminum dust administered to rabbits prevented silica dust from dissolving into the silicate acid that damaged the lungs. A short study on volunteers was conducted in 1939, but some medical professionals wanted more human testing (with control groups) done. Unfortunately, the long delay needed for results of these tests combined with the immediate need of the mines meant that further testing was never conducted, and many mines adopted the powder wholesale without knowing its full effects.[63] The Canadian government neither endorsed nor condemned the practice of administering "dust therapy" in the absence of research showing either its benefits or its risks. McIntyre Powder was eventually used in America, Australia, Chile, Mexico, and South Africa as well as in Canada.

Starting shortly after McIntyre Powder's introduction to international markets in the early 1940s, international research on aluminum workers and criticism from bodies such as the American Medical Association cast doubt on its harmlessness.[64] This critique combined with increasing resistance from workers forced to take McIntyre Powder treatments against their will.[65] The industry's response to these critiques is telling. Indeed, as much as the McIntyre Foundation claimed to be an independent not-for-profit research body interested only in the health of miners, evidence shows significant overlaps between Ontario mining companies and McIntyre Foundation governance. A 1955 list of the foundation's directors, for example, is dominated by the names of presidents and directors of major Porcupine mines.[66] The foundation became an important tool for forwarding corporate interests. Its main concern was the continuation of extraction in Ontario without the need to engage in expensive and difficult change. The foundation's answer to workers' complaints included painting change houses gray so that the dust would not show up as much, attempting to manufacture white powder, and aggressive "educational" campaigns broadcasting McIntyre's benefits.[67] Although the foundation claimed that its powder should always be used in conjunction with other dust-control methods, in practice it lent its weight to opposing the introduction of threshold-limit values for silica dust and apologized for the companies that could never "effectively and practically control dust to the point where it does not constitute a hazard for the men."[68] The international community finally blew the whistle on McIntyre Powder. Citing Australian research that showed aluminum powder heightened tuberculosis risk, the United Steelworkers of America and other North American unions began rejecting the forced inhalation of the powder in miners' change rooms in the 1950s. The practice of administering the dust to miners finally ended in all mines in 1979.[69]

We still do not know the precise impacts of McIntyre Powder on human bodies, but evidence shows that it lacked either the curative or the preventative powers claimed by the industry. In Canada medical research on the links between aluminum powder and neurological disease is ongoing. This work is thanks largely to the efforts of Janice Martell, who connected McIntyre Powder to the Parkinson's diagnoses of her father, Jim Hobbs. Over the past decade, she helped organize the submission of ninety-two claims for compensation to the Ontario Workplace Safety and Insurance Board and has collected the names of hundreds of miners exposed to McIntyre Powder during their working lives, a disproportionate number of whom now suffer from various neurological symptoms.[70] Globally, the precise health impacts of the half-century campaign to eradicate silicosis with aluminum powder are even more difficult to calculate. In Mexico in 1956, the foundation declared 1.25 million treatments given. In Australia 2,337 miners, or 72 percent of Western Australia's total underground workforce, took McIntyre Powder by 1958. With little or no documentation of workers treated in these international fields, the precise impacts of McIntyre Powder are impossible to trace.

McIntyre Powder embodies the international character of Porcupine mining, the hegemony of technological solutions to environmental problems, and numerous other themes in the history of the industry in Canada such as its adopted socioracial hierarchies and gradually escalating scale. In searching for a technical solution to silicosis, Ontario mines would once again offload the cost of overcoming environmental problems elsewhere. When silicosis numbers began to decline after the Second World War, the McIntyre Research Foundation took credit, even though contemporaneous advances in ventilation and dust-control technology likely drove the decline.[71] But neither the disease nor its unwanted "cures" went away. At the Elliot Lake uranium mines, miners went on strike in 1974 after high cancer rates among workers alerted unions to the inadequacy of current health measures. Workers rejected aluminum powder as a solution and as part of their negotiations with the employer made aluminum therapy optional instead of compulsory. The McIntyre Foundation would continue to assert McIntyre's harmlessness into the 1990s, despite the cancellation of most of their contracts, several high-profile exposés, and widespread public critique.[72] Although miners had always disliked it, the full impacts of aluminum on their bodies and brains are still unknown. Through the McIntyre Research Foundation's work, the mining industry was able to push the burden of silicosis into the far future. After more than a century of violence against workers' bodies, both silicosis and its "cure" have proved the slowest and deadliest environmental disaster stemming from the mining industry at Porcupine.

Science, Survival, and Silicosis in the 1920s

In the aftermath of the Hollinger fire, Supreme Court justice T. E. Godson wrote, "The evidence does not indicate, nor do I find that a recognized danger was carelessly cast aside in order to achieve major production."[73] The same might be said of silicosis. As tempting as it may be, the silicosis crisis cannot be used as evidence of a malicious corporate scheme to undermine the health of miners. Both silicosis and the Hollinger fire snuck up on mine managers and miners alike, who acted to look after their interests. Silicosis was recognized because of the work of international science and the rise of consciousness around industrial disease and workers' compensation at home and abroad. But it was also a natural product of the rock that had always been there and had always released silica dust. These processes were only ever partially under human control. The increased tempo of industrial mining produced waste above and beyond what the mines had ever imagined dealing with. As the waste and its toxins built up on the surface, below the ground, and in miner's bodies, it became increasingly difficult to keep track of. When it became clear to the mining companies that it represented an existential threat to their existence, they reacted with all the tools at their disposal.

When silicosis threatened the mining industry in Canada, mining's benefactors turned to the same solutions that had worked well for them in the past: by drawing on international expertise for technical solutions. They understood that there would be a financial impact, so the goal became minimizing those costs. McIntyre Powder, like the Mattagami dam, offloaded part of the costs of overcoming an environmental problem to someone else—in this case, miners' bodies. As the mines forged ahead into ever-deeper shafts and lower grades, mining companies proved perfectly capable of looking ahead to avoid environmental crises from adversely affecting profits. McIntyre Powder proved especially efficient in this regard, because the effects of both silicosis and aluminum therapy were hard for miners to detect on their own until it was too late. As long as they controlled the science, mining companies could claim that their actions were both legitimate and benevolent.

American silicosis historians David Rosner and Gerald Markowitz read silicosis's story as the first instance in which people finally began to doubt the notion that science and technology were unvarnished forces for good. The dust's clear link to industrial extraction and the inability of the mining companies to adequately address it, they argue, shook human faith in modern projects of industrial development.[74] However, the evidence from Canada shows that no one north of the border ever doubted technical solutions in the 1920s. There was an unwavering faith at all levels of the documentary evidence in the possibility of international or technical expertise to solve environmental problems. When the international scientific community failed to provide satisfactory answers, mine management looked to do *more* science, not less, and created McIntyre Powder. It was the same faith in technical environmental management that shaped the nascent conservation movement in Porcupine in the early 1920s, while never questioning the legitimacy of the mining industry to fill lakes with tailings.

The McIntyre Foundation and the endurance of its assertions of harmlessness are a neat microcosm of the larger attitude of the industry toward the consequences of environmental change. In McIntyre Powder, the industry found a perfect vessel for transferring the costs of extraction. The complexity of lung disease and its slowness rendered silicosis invisible, to the point that the industry could claim they had solved it without fear of repercussion if their claims proved unfounded. Over the years, this strategy largely proved effective. Canadian mining companies continued to use many of the same strategies for this new challenge that they had used to confront barriers to production since the earliest days of the Porcupine rush. The difference is that now the immense scale of the distances, inequalities, and technologies involved rendered the project much bigger.

Conclusion

Industrial Dreams, Industrial Nightmares

Despite regular setbacks related to ongoing power and labor shortages, the story of 1920s Canadian gold mining was one of ascendance. Buoyed by postwar demand, the entire sector doubled its production between 1919 and 1929. Porcupine gold was at the center of this success story as industrial hard-rock mining techniques opened up new possibilities for profit in the north. In 1920, Porcupine produced over 90 percent of Ontario's gold.[1] During the peak years of production between 1925 and 1929, it produced more than half (and up to two-thirds) of Canada's total gold output annually.[2] More important, it laid the scientific and industrial infrastructure for the extension of industry along the Abitibi gold belt from Kirkland Lake to Val-d'Or and beyond. As Porcupine's production declined, its successors carried on its legacy and ensured the Canadian gold industry would continue until the 1970s, with a brief slowdown during the Second World War.

As part of its newfound productivity, Canada began to develop a reputation for international excellence in specific areas. Leadership was forged especially in environmental health and safety, stemming from the industry's experience in the Hollinger fire and the subsequent silicosis crisis. Those in positions of power in the industry framed waste and dust problems as problems of management, and thus perfectly fixable given sufficient application of science and technology. In both cases, it was Canadians' unfettered optimism more than their technological prowess that proved most irresistible to international audiences looking for solutions to intractable industrial problems. Canadian mining's advocates repeatedly asserted that they could have it all: profitability, healthy workers, and unspoiled environments—despite mounting evidence to the contrary accumulating on the land and in workers' bodies.

As I write this, gold continues to come out of Porcupine year after year. Most gold mining frontiers are plagued by boom and bust, and their histories offer disjointed glimpses of the industry at different points in its evolution. Porcupine is not like that. As a result, the area presents a fluctuating but continuous picture of the evolution of Canadian mining. Although Porcupine shares many characteristics with other mining zones, there are few other places in the world that have produced gold without pause for more than a hundred years.[3] In an industry famed for instability, uncertainty, and unsustainability, Porcupine proved durable. Contrary to the boom-bust critique of extraction as a short-term and shortsighted endeavor benefiting a handful of foreign investors at the expense of local interests, mining at Porcupine endured within limits imposed from both within and without the North. As the land changed around them, people at Porcupine changed with it. Pits and tailings piles became accepted and even celebrated parts of life in the North because they signified progress for an industry plagued by setbacks.[4] In a world where historians point to mining as the emblem of our unsustainable relationship with nature, Porcupine looks like evidence that given the right timing, the right networks, and the right geology, extraction can provide some semblance of economic and social stability.

What is Porcupine's secret? If there is one thing that sets Porcupine apart, it is its timing. As Porcupine turned toward an extractive future, miners, mining companies, and legislators could look back on the nineteenth-century world and see enormous potential profits, dangerous social and legal pitfalls, and some clear lessons for its own development. In the early twentieth century, northern Canada was on the cusp of industrial transformation and increasingly connected to the global market economy. Fifty years of mining experiments around the world offered the intellectual, technological, and economic tools required to make Porcupine mining successful. Almost the entire story of industrial mining ingenuity lay ahead of the industry when gold was discovered in 1909, including a host of techniques for extracting ever more finely distributed fragments of hard-rock gold. At Porcupine, that gold happened to exist in sufficient quantity to support the industry, even when outside factors threatened to bring it down. Of all the world's gold mines, Porcupine's relatively recent origin meant that it could take advantage of the most recent hard-rock mining innovations without being burdened by old technological infrastructures. Unlike its counterparts in South Africa and the American West that had witnessed the gamut of mining progress since the late nineteenth century, Porcupine was a sort of extractive tabula rasa. Porcupine also benefited from the support of American capital and renewed demand for metals in the postwar era. Finished with its speculative phase and comparatively free of the labor politics that bogged down production in other fields, it enjoyed a reputation as a stable, well-understood investment option. The result was that by the end of 1930, Porcupine would intervene in an international conversation about the science and safety issues in gold mines around the world. Then as now, extraction at Porcupine persisted in the Abitibi because it could draw on powerful corporate, government, and human networks. As a result, Porcupine became a birthplace and a testing ground for modern relationships with the environment in Canada and in the mining community broadly.

Porcupine's environmental history cannot be fully explained without referring to both the internal and the external factors that drove successive crises and their solutions over time. International capital, markets, and technologies enabled industrial mining in northern Ontario, but only in combination with an enthusiastic provincial government and preexisting Indigenous infrastructures. A wildfire and a world war worked together with the facts of geology to encourage the consolidation of smaller properties into the extractive giants that would come to dominate the Porcupine for the next century. The large-scale industrial mines operated under the combined power of both international labor and northern waterpower. While the gold they extracted flowed to American mints and then all over the world, the wastes from production accumulated locally. When some of that waste exploded during the Hollinger fire, rescue came from outside of Canadian borders. Similarly, when Porcupine dust proved dangerous to miners, silicosis science flowed from all over the world to help Canadians make sense of the crisis. Canadian expertise also flowed outward, with new safety measures for gold mines (following the Hollinger fire) and new industrial disease cures (McIntyre Powder). This book shows that mined landscapes are the product of the convergence of local and international forces. While much of the world's mining history has been written at the regional or national scale, it makes sense to consider how these places connect more broadly across borders. More than a resource reservoir for Ontario, Porcupine made Canada a "treasure house to the world."[5]

In the end, though, neither technology nor optimism could overcome the fact that gold is finite, and without new discoveries the industry would die. The specter of exhaustion always haunted Porcupine, as it does all gold mines. Even as they pulled back the muskeg to expose massive veins of gold-bearing rock, prospectors, miners, investors, geologists, and legislators all understood that one day it would all be gone.

This history of Porcupine after 1909 can be understood as a series of successive attempts to push back that inevitability. The only difference between Porcupine and other places is that Porcupine succeeded for longer, largely because of its timing and the nature of its deposit, which was larger and deeper than any of its first discoverers could have imagined. Despite these advantages, Porcupine shows how the effort to keep mines alive required displacing Indigenous people, exploiting working people, and profoundly damaging the environment. Not everyone bore the weight of gold equally.

The next century of mining in northern Ontario built on the foundation set deep in Abitibi bedrock between 1909 and 1929. As extraction entered its third decade on the land, some of the instability that had plagued the early years faded. The industry experienced a new confidence, buoyed by its improving fortunes and reinforced by its successes. The period was marked by expansion, celebration, and reflection by the companies and the government. Many of the prospectors who entered new goldfields at Red Lake and Woman Lake in 1925 and 1926 did so with technology, capital, and experience gained at Porcupine. New rushes spawned new human and environmental challenges that in turn went on to influence the next wave of Canadian extraction. Tentative experiments of the 1920s expanded into full-scale expansion projects in the 1930s. After recovering from a mill fire in 1929, Dome began looking overseas to South African frontiers and obtained stocks in several South African properties through the 1930s. The company also sent prospectors into Quebec and the Northwest Territories.[6] Similarly, Hollinger pursued a "policy of exploration and investigation of outside properties. . .more vigorously than in former years," especially at nearby properties around Kirkland Lake.[7] With these extensions into local and international frontiers, Porcupine secured its legacy.

By 1930 the mines were surrounded by a community that drew its identity from the industry. These years became an important moment for storytelling about people and their relationship to the land as it had evolved since 1909. This became an especially urgent project as the first generation of Porcupine miners died.[8] The *Porcupine Advance* told Hollinger's foundation story in 1929, and Hollinger produced a movie (shown at the Goldfields Theatre in Timmins) portraying the underground workings and other features of the mine.[9] The *Porcupine Advance* published a regular "Ten Years Ago in Timmins" segment and ran a history of the origin of area names in 1930.[10] In 1936 the general manager of

McIntyre Mine, R. J. Ennis, published a history of the mine, and Stephen L'African wrote an exhaustive piece on "early mining facts of Porcupine."[11] Despite the fact that industrial mining possessed few of the characteristics of early prospecting, these accounts invoked an older narrative of independence, adventure, and risk. The 1930s marked the sensational return of the "good old days" of the original Porcupine gold rush.[12] Such stories served both as a marker for the technological advancements made since 1909 and as a way of romanticizing what had become corporate-dominated industrial work.

These stories were not constrained to Canada. International media also wrote about Porcupine as part of a bigger (romantic) era of gold discoveries. For example, Kalgoorlie mining investor and prospector James P. Hallahan recounted his adventures at multiple goldfields, from Chile to New Zealand to Mexico to Australia to the United States and finally to Canada in the nineteenth century, and a full page of the *Australian* was devoted to "the importance of Ottawa" in the "primary industries of the Empire."[13] The celebration of Porcupine was tied to celebrations of gold mining generally and to the accomplishments the industry had experienced across multiple goldfields by the 1930s.

Among these narratives, a sense of Canadian exceptionalism on the international mining economy was taking shape. In a 1929 article called "Toronto Becomes Big Mining Centre," *Canadian Mining World* editor Sidney Norman declared the Toronto Stock Exchange "not only the leading mining market on the North American Continent. . .but the greatest in the world." Norman cited Toronto's unique location at the "centre of the most productive and intensely industrialized area in the Dominion," only a short train journey away from the location of "one of the greatest mining movements the world has witnessed."[14] The *Globe* said that, thanks to northern Ontario, "Canada has today earned a position of international prominence among the mineral producing countries in the world." Several sweeping accounts centered mining in Canadian economic history. They chronicled extractive ventures from coast to coast and romanticized the visionary men who had overseen their discovery and development.[15] In 1935 the Canadian Department of Mines published *Gold in Canada,* a complete report on the state of the industry. It began with an analysis of Canada in relation to the world, comparing Canadian production against contemporary Australian, Russian, American, and South African figures.[16] It told the story of gold from ancient Egypt to the present and identified distinct eras of the industry: the conquest of

In 1936 the Department of Mines and Technical Surveys produced a series of photographs of the Porcupine mines highlighting the technology and infrastructure of the camp in a classic example of industrial sublime. "Hollinger Assay Office—Pouring pots containing molten lead & slag into a tray in front of row of muffle furnaces. Timmins, Ont., 1936," PA-017587, MIKAN no. 3374948, Canada Department of Mines and Technical Surveys, Library and Archives Canada, Ottawa.

central and South America, the nineteenth-century gold rushes exemplified by Australia and California, and the industrial period dominated by South Africa and Canada. Above all, *Gold in Canada* suggested that the greatest parts of Porcupine history were yet to come.

In 1936 the Canadian government's Motion Picture Bureau collaborated with the Department of Mines and Resources to create a series of films depicting gold mining in Porcupine, Kirkland Lake, and Norada.[17] The Department of Mines and Technical Surveys also produced a series of photographs of Porcupine mines, highlighting the technology and infrastructure of the camp. Classic examples of industrial sublime, these images portray massive cyanide vats, mills, offices, furnaces, shafts, mine timbers, trams, nuggets, and gold bricks. The people in these images are seamless parts of the mining system: miners are pictured moving between parts of the mine, drillers and drill bits lean against the walls of huge underground chambers, long rows of ball mills gleam under the eves of enormous structures, and assayers carefully pour molten metals. In these images, Porcupine, its geology, and its environment become a backdrop for displaying northern Ontario's industrial ascendance.

Today, more than half of the world's mining operations are Canadian owned, equipped, or engineered. Two-thirds of all the world's mining companies headquarter in Canada to take advantage of favorable tax regimes, world-renowned expertise, and supportive organizations (including the Prospectors and Developer's Association, whose Toronto conference is the premier international event of its kind, attracting twenty-five thousand people from 135 countries every year). Although Canadians continue to be active extractors of mineral wealth within their own borders, two-thirds of the value of Canadian-owned assets are held outside of Canada. Most international Canadian mining activity occurs in Latin America, especially Mexico and Chile, but Canadians also assert considerable influence in Africa, Southeast Asia, eastern Europe, and the United States.[18]

As a case study, Porcupine represents an early foray in what became Canada's regular habit of extractive interventionism on behalf of Canadian-owned mining companies, often with disastrous human and environmental consequences.[19] Many of these companies can trace their lineages to the Porcupine. Then as now, the combination of overbearing optimism and aggressively wielded scientific authority downplayed the precautionary inclinations of people living and working on mined landscapes. The gaping chasm between Canada's asserted moralism as a human rights and environmental leader versus its actual practices in places like South Africa and Latin America has its roots in the bedrock of the Canadian Shield. While I was in the archives during the first stage of my research for this book, a coalition of Indigenous Maya Q'eqchi' women were busy suing Canadian-owned Hudbay minerals in Ontario's Supreme Court for human-rights violations related to the appropriation of traditional territory. Around the same time, Barrick Gold's Pascua-Lama operation was facing a $16 million fine levied by the Chilean government for environmental damage.[20] As I sat down to write two years later, Imperial Metals managed to avoid fines after it spilled ten million cubic meters of arsenic and selenium-laden tailings into Quesnel Lake, in British Columbia.[21] As I sent the manuscript out for review for the first time in 2019, thousands of Turkish protesters spoke out against the Canadian-owned Alamos Gold's Kirazli mine and its impact on the local environment.[22]

Modern controversies around Canadian extraction are echoes of Canada's long struggle with its own intractable environment, its crippling insecurities about its position on the world stage, and its discovery that obscuring the true costs of mining could grant astonishing rhetorical power. The silicosis crisis particularly exemplifies the way that the costs of industrial extraction could be transferred by mining companies to other people and future generations under the combined forces of science, rhetoric, and structural inequality. The same technological optimism that granted McIntyre Powder such longevity despite little enthusiasm on behalf of either miners or regulators can be heard echoing in modern industry rhetoric. Canadian mining companies wield their benevolence like a club.

Direct historical lines of cause and effect cannot easily be drawn between two decades of mining at Porcupine and the modern power and influence of Canadian mining companies on landscapes around the world. Yet it is equally clear that echoes of history reverberate into the present. The confidence of the expanding mining companies in the 1920s, the rise of the industrial sublime, and the stories Canadians started telling themselves about their extractive pasts all had a role to play in securing Canada's place in the highest echelons of the international mining industry well into the twenty-first century.[23] It was during these early years where Canadians found their place at the table. The dominance of mining in the North, the insistence on technological solutions to environmental problems, and the redistribution of the costs of industrial mining to workers and Indigenous people became recurring patterns across time and space.

To this day, Canadian mining companies actively promote their activities as an extension of the gold frontier, drawing on a white-settler imaginary to frame themselves as helpers or rescuers of "developing" economies in under- or inefficiently exploited mineral-bearing landscapes. Some scholars see the 1980s "Washington consensus" and the subsequent neoliberal restructuring of the "Global South" as a critical moment explaining the twenty-first-century influence of the modern Canadian industry.[24] During this time, World Bank–led structural adjustment programs moved resource-rich states toward neoliberalism, while the Canadian government provided export credits to mining companies and development assistance for policy reforms friendly to extraction. Canada actively cultivated its reputation as a global leader in specific areas, especially mine financing and exploration. The powerful combination of external opportunity and internal enthusiasm, observers have argued, created exactly the right conditions for the rise of Canadian mineral imperialism in many parts of the "developing" world.[25]

Canadian enthusiasm and ability to respond to the events of the 1980s were not inevitable; rather, they were the result of a long series of human choices within particular physical environments. The focus on financing, exploration, and (more recently) green technology was not incidental or natural. At the same time, Canada's rise on the international stage cannot be traced to a single historical moment or event. Understanding Canada's modern role as an imperial extractive power means recognizing its long and cumulative history of interactions with other mining landscapes around the world. By zooming in on northern Ontario, an early and influential location of Canadian experiments with industrial gold mining, we can identify some of the formative foundational layers that went into the construction of a modern empire.

Understanding Canada's mining history means looking both within and outside of the state. This book has sought to tackle both perspectives by thinking about Porcupine as a "node" or a temporary metropole for the forces of international mining during a sensitive time in the industry's development. Events in the Abitibi cannot fully explain the entirety of the complex behemoth that is the modern Canadian mining industry. What it can do is offer some key lessons to help make sense of the contemporary Canadian mining. The Porcupine gold rush and its aftermath shows the extents the industry was willing to go to overcome its environmental limits. Relentless optimism combined with sufficient capital produced technological solutions. At Porcupine technological solutions worked best when their costs could be transferred to places where they were not quantified.[26] The alienation of Indigenous communities from Abitibi land was the first step in this process. Later, working bodies (including *future* human bodies) and landscapes would function as repositories for mining's harms during the Porcupine fire, the Mattagami River Power Project, the Hollinger fire, and the silicosis crisis. If there is something especially insidious about Canadian mining from Porcupine to the present, it is its unearned optimism in the face of these costs. So long as mining occurs in a context where there are discrepancies in power between people, it will exacerbate inequality through the distribution of harm on those with little immediate recourse.

But northern Ontario's history also teaches us that there is power in connection. In the preceding chapters, I have described how bureaucrats, companies, and scientists worked across borders on mining projects, gaining power and momentum over time. Missing from this list of international actors past and present are the individual communities that bear the burdens of extraction on the land and in their bodies. Communities have been less effective than corporations at looking across borders for answers to their problems, environmental or otherwise. The reasons are complicated. Structures of race and power mean that working-class white-settler communities in northern Canada do not see themselves as holding common cause with Indigenous Maya Q'eqchi' communities in Guatemala and vice versa. And indeed, the issues they face look substantively different because of the structural inequalities that grant mining companies greater license in places like Guatemala than they do in northern Canada. Nevertheless, mining communities are all related, even if they did not ask to be. Their relationship is based on political, economic, and geographic processes beyond their control, and thus difficult to grasp. Isolation benefits extractors because isolated communities cannot trace the costs of gold mining across time and space. Only when the pieces of mining history are held up next to each other do they tell a complete story. If the people who live with gold mining can see themselves as part of a larger network of communities dealing with shared injustices, they may be better equipped to find solutions.

Ultimately, mining's legacy for the people and the environment in northern Ontario will be difficult to evaluate. The mobility of miners means we will never know the total damage to their bodies wrought by silicosis (or by its aluminum-based "cure," McIntyre Powder) in Ontario or around the world. The economic and social losses experienced by Mattagami First Nation in the wake of 1920s dam building are difficult to quantify given the multifaceted and multigenerational effect of colonial violence. Even the basic financial costs to Canada as a whole cannot be adequately calculated. Cleanup at nearby Kam Kotia mine has already cost more than $75 million (original estimates were for four years at a cost of $27 million in 1999).[27] The continued life of the industry was ensured only through enormous sacrifices on behalf of miners, Indigenous people, community members, and all their future generations. As mining industrialized, companies could never fully shed the inherent risk and uncertainty of their business. They could only forestall, not eliminate, environmental and geological impacts of their activities on the land. In the twenty-first century, we have inherited the consequences of deferring these impacts in our societies and in our bodies. As the scale of extraction continues to expand, its harms become more difficult to defer.

Epilogue

Living Well with Mined Land

We continue to live with the legacy of Porcupine at thousands of mine sites in Canada and around the world. I first learned to be curious about the way gold mining changed the land because I lived with its consequences, but until recently those consequences had always originated in the past. Midway through my research for this book, the arrival of a modern mining regime in my hometown of Wells, British Columbia, showed me the power of a replicating pattern I had begun to recognize. I could see it clearly as I looked from Wells to Porcupine and back again. When a new gold mining company began the environmental assessment process and met with my neighbors, it made the same promises mining companies have been making for a hundred years to people around the world: jobs, wealth, and safe environments. The archives were full of evidence that made me question those promises, but a glance across the river at the orange tailings from the 1930s Cariboo Gold Quartz mine was evidence enough. As I read about the Digby suit in Porcupine, the new exploration work occurring in Wells resulted in federal and provincial fees for noncompliance, including violations for failure to properly treat wastes. As I traced the international careers of Porcupine managers and geologists, I could not help but notice how my local mining company's executive included mining men with résumés extending across North and South America. Even as I worked to understand how local and extralocal forces had transformed the land in northern Ontario at the turn of the twentieth century, I watched those same forces at work across the valley from my childhood home in the twenty-first.

The connection between Wells and Porcupine seemed as obvious as it was difficult to articulate. Because I swam in a river contaminated by heavy metal pollution and rode my bike on a mudflat laced with cyanide, I held mining history physically in my body in a way that could not be expressed in words. Gold rushes are famous for their boom-and-bust cycles that leave ghost towns in their wake. Hope, triumph, and tragedy are essential elements of all good mining stories. But their unpredictability and rhetorical power also make them hard to write about. Even as I worked to pin the history of gold mining to the page in some kind of comprehensible order, it kept piling up around me. I felt paralyzed in the face of a rhetoric that insisted community concerns were a symptom of poor education rather than legitimate fears based on more than a century of historical precedent.

Through it all, the contaminated tailings sat silently in the valley bottom across from my home. The promise of prosperity, longevity, and health told by mining's advocates did not fit with the story those tailings told. Gold rushes had already come and gone twice, and neither time had ended well. We did not need the potential benefits and consequences of modern mining explained to us. We had several thousand tons of dislocated, mercury-laden, acidified orange dirt to tell us everything we needed to know. Yet without a hint of irony, we were assured that the solution to the problems created by gold mining was *more gold mining*. After the sixteen-year life span of the mine was exhausted, some of the Cariboo Gold Quartz contaminated tailings would be cleaned up as part of the remediation plan. Perversely, our contamination became evidence that gold mining was appropriate and that it could fix the errors of history. Because Wells was already damaged, the impacts of mining were seen as less consequential.

The Weight of Gold is my effort to put the story told by tailings piles into words. That story carries little weight in its current form in the decision-making process engaged by the province, the community, and the mining company. In the end, we are being asked, in the face of the evidence, to trust that this time the results of mining will be different. Maybe it will, maybe it will not, but my money is on the latter. As the climate changes, the costs of extraction get increasingly difficult to defer, and more of us get asked every day to accept life in damaged spaces. Decisions about gold mines and all other natural resource projects must broaden their definition of "impacts" and "benefits" to include both historic harm and the well-being of future generations. I hope that one day the story that the land tells us holds as much weight as the stories we tell ourselves and that communities can make decisions about their own futures with a clear understanding of their true costs.

In the meantime, we must learn to live well with mined land—and not just mined land, but all kinds of land damaged and altered by anthropocentric climate change. Doing so means extending our ability to value and care for new kinds of spaces. No longer is it sufficient to possessively guard those places designated pristine and untouched. Such an ethic is harmful to the people who have touched and managed that land since time immemorial, to land itself that sometimes seeks (mindful) disturbance to thrive, and to all those other spaces that are equally deserving of our care. Sorting our landscapes into categories of worthy and unworthy of stewardship deprives us of the potential for a life connected to many kinds of places, including the ones that lie just outside our windows. An understanding of a place's history, including toxic, traumatic history, should enrich our connection to it, not invalidate it. Mined places deserve careful handling because of their pasts, not in spite of them.

Bibliography

Archival Sources

Archives of Ontario, Toronto

Dome Mines Company Fonds. F 1350.
H. C. Peters Fonds. C 321.
Hollinger Company Fonds. F 1335.

Library and Archives Canada, Ottawa

Department of Health Fonds. RG 29.
Department of Indian Affairs and Northern Development Fonds. RG 10.
Department of the Secretary of the State Fonds. General Correspondence. Mrs. Mary Coutts and others—Enquiries regarding the porcupine fire. RG 6 A-1 Vol 150.
Earnest F. Pullen Fonds. MG 30, E 219.
Mineral Resources Branch Sous Fonds. RG 87.
Northern Canada Power Co. et al. v. Hollinger Consolidated Gold Mines et al. P. C. 1924. MG 28-III35, box 57, Library and Archives Canada, Ottawa.
William S. Edwards Fonds. MG 30.

Queen's University Archives, Kingston, Ontario

School of Mining and Agriculture Fonds

William Ready Divisions of Archives and Research Collections, McMaster University, Hamilton, Ontario, Canada

Porcupine Postcard Collection

Digital Newspaper Archives

NATIONAL ARCHIVES OF AUSTRALIA, TROVE

Adelong and Murrumbidgee District Advisor (New South Wales)
Advocate (Burnie, Tasmania)
Age (Melbourne, Victoria)
Armidale Express and New England General Advertiser (New South Wales)
Australasian (Melbourne, Victoria)
Barrier Miner (Broken Hill, New South Wales)
Bendigo Advisor (Victoria)
Clarence and Richmond Examiner (Grafton, New South Wales)
Daily Advertiser (Wagga Wagga, New South Wales)
Daily Herald (Adelaide)
Daily News (Perth, Western Australia)

Daily Post (Hobart, Tasmania)
Darling Downs Gazette (Queensland)
Evening News (Sydney)
Geelong Advertiser (Victoria)
Goulburn Evening Penny Post (New South Wales)
Gungandai Times and Tumut, Adelong, and Murrumbidgee District Advertiser (New South Wales)
Gympie Times and Mary River Mining Gazette (Queensland)
Mail (Adelaide, Southern Australia)
Mainland Daily Mercury (New South Wales)
Maitland Daily Mercury (New South Wales)
Northern Star (Lismore, New South Wales)
North West Post (Formby, Tasmania)
Riverine Herald (Echuca, Victoria, and Moama, New South Wales)
Sun (Sydney, New South Wales)
World's News (Sydney, New South Wales)

NATIONAL ARCHIVES OF NEW ZEALAND, PAPERS PAST

Akaroa Mail and Banks Peninsula Advisor (Canterbury)
Auckland Star (Auckland)
Colonist (Nelson)
Daily Times (Otago)
Dominion (Wellington)
Grey River Argus (West Coast)
Hawera & Normandy Star (Taranaki)
New Zealand Herald (Auckland)
Press (Canterbury)
Star (Christchurch)
Stratford Evening Post (Taranaki)
Wanganui Herald (Manawatu-Wanganui)
Woodville Examiner (Manawatu-Wanganui)

PROQUEST HISTORIC NEWSPAPERS

Globe/Globe and Mail (Toronto)
New York Times (New York)
Wall Street Journal (United States)

UNIVERSITY OF CALIFORNIA, CALIFORNIA DIGITAL NEWSPAPER COLLECTION

Los Angeles Herald (Los Angeles)
San Francisco Call (San Francisco)

UNIVERSITY OF VICTORIA, *DAILY COLONIST* DIGITAL COLLECTION

Daily Colonist (Victoria)

Government Documents

Annual Report of the Ontario Bureau of Mines, 1891–1935. Toronto: King's Printer, 1892–1936.

Canada. Statistics Canada. 1911 Census. Ottawa, 1911.

Dominion of Canada Report of the Department of Mines and Resources for the Fiscal Year Ended March 31, 1937. Ottawa: King's Printer, 1937.

Gibson, Thomas. *The Mining Laws of Ontario and the Department of Mines.* King's Printer: Toronto, 1933.

Government of Canada. "Ottawa Data." In "Almanac Averages and Extremes: Historical Climate Data." Accessed May 18, 2017. http://climate.weather.gc.ca/.

Grisdale, J. H. "Report of the Director." In *Annual Report of the Experimental Farms for the Year Ending March 31, 1912.* Ottawa: King's Printer, 1912.

Harrington, Daniel. "Progress in Metal-Mine Ventilation in 1930." *Circular 6469, Information Circular Department of Commerce, (United States) Bureau of Mines.* Washington, DC: US Department of the Interior, July 1931.

———. "Progress in Mine Ventilation." *Circular 6136, Information Circular Department of Commerce, (United States) Bureau of Mines.* Washington, DC: US Department of the Interior, May 1928.

———. "Work of the Safety Division of the United States Bureau of Mines, Fiscal Year 1930." *Circular 6400, Information Circular Department of Commerce, (United States) Bureau of Mines.* Washington, DC: US Department of the Interior, November, 1930.

Hearings Before the Committee on Mines and Mining United States Senate Seventeenth Congress Second Session on S. 2079. "A Bill Authorizing an Appropriation for Mining Experiment Stations of the United States Bureau of Mines, Part 2, 15 February 1929." Washington, DC: Government Printing Office, 1929.

Macoun, W. T. "Report of the Dominion Horticulturalist." Second Session of the Twelfth Parliament of the Dominion of Canada Session 1912–13. No. 16, March 31, 1912. Sessional Papers, vol. 9.

Price, S. "Report re. Limitation of the Hours of Labor of Underground Workmen in the Mines of Ontario." Ontario, Sessional Papers, No. 85, 1913.

Report on Mining Accidents in Ontario. Bulletin No. 9 (1911)—Bulletin No. 75 (1930). Toronto: King's Printer, 1912–31.

Media

Armstrong, Diane. "Remembering the Porcupine Fire." *Timmins Daily Press,* July 25, 2012. .

Bachman, Karen. "Pioneers Share Fire Tales." *Timmins Daily Press,* July 8, 2011.

"Barrick Third Quarter 2017 Results." October 25, 2017. http://www.barrick.com/investors/news/news-details/2017/Barrick-Reports-Third-Quarter-2017-Results/default.aspx.

"Chile Fines Barrick Gold 16m for Pascua-Lama Mine." BBC News, May 24, 2013. http://www.bbc.com/news/world-latin-america-22663432.

"Environmental Cleanup Continues at 'Canada's Worst Mining Disaster' in Timmins." CTV News Northern Ontario, June 8, 2017. https://northernontario.ctvnews.ca/environmental-cleanup-continues-at-canada-s-worst-mining-disaster-in-timmins-1.3449983.

Goldcorp. "Goldcorp Signs Important Resource Development Agreement with Four First Nation Communities in Timmins, Ontario." News release, November 24, 2014. https://s24.q4cdn.com/382246808/files/doc_downloads/goldcorp_archive/0_0_goldcorp_csr_2016_full.pdf.

Goldcorp. "Responsible Mining," 2016 Sustainability Report. Accessed November 15, 2017. http://csr.goldcorp.com/2016/communities/stakeholder-engagement.

———. "Porcupine Project." Accessed April 1, 2018. https://www.goldcorp.com/English/portfolio/operations/porcupine/default.aspx.

Ontario Ministry of Northern Development and Mines. "Metallic Minerals." May 25, 2012. http://www.mndm.gov.on.ca/en/mines-and-minerals/geoscience/metallic-minerals.

Print Magazines and Print Newspapers

Canadian Mining Journal (1908–23)
Engineering and Mining Journal (1909–23)
Globe and Mail
Labor Gazette (1913–30)
Northern Miner (1910–28)
Porcupine Advance (1912–29)
Toronto Star

Published Primary Sources

"Aluminum in the Prevention and Treatment of Silicosis." *Journal of the American Medical Association* 130, no. 17 (April 1946): 1223.

Davis, Harold Palmer. *The Davis Handbook of the Porcupine Gold District.* New York: n.p., 1911.

Elliot, Jabez. "Silicosis in Ontario Gold Miners." *Canadian Medical Association Journal* (1924): 930–37.

Goralewski, G. "Clinical and Animal Experimental Studies and the Question of Aluminum Dusty Lung." *Archiv fur Gewerbepathologie and Gewerbehygiene* 9 (1939): 676–88.

Gray, John. "The Fire That Wiped Out Porcupine." *Maclean's*, February 1, 1954, 20–38.

Oliver, Thomas. "A Discussion of Miners' Phthisis." *British Medical Journal* 2, no. 2228 (1903): 568–73.

Prusila, George. "Northern Ontario." Interview by George Peck, May 1982. Transcribed by Wendy Mayhew. *Memories and Music,* Summer Canada Project Oral History Program, Sudbury Public Library.

Riddell, A. R. "A Case of Silicosis with Autopsy." *Canadian Medical Association Journal* 15, no. 8 (1925): 839–41.

Robinson, A. H. A. *Gold in Canada.* Ottawa: King's Printer, 1935.

"Silicosis: Records of the International Conference Held at Johannesburg, 13–17 August 1930." International Labor Office (League of Nations): Studies and Reports Series F (Industrial Hygiene), no. 13. London: P. S. King & Son, 1930.

Trudeau, Francis B. "The Objectives and Achievements of the McIntyre Research Foundation." *American Medical Association Archives of Industrial Health* 12, no. 1 (1955): 2–3.

Secondary Sources

Abel, Kerry. *Changing Places: History, Community, and Identity in Northeastern Ontario.* Kingston: McGill-Queens University Press, 2006.

Ally, Russell. *Gold & Empire: The Bank of England and South Africa's Gold Producers, 1886–1926.* Johannesburg: Witwatersrand University Press, 1994.

Andrews, Thomas. *Coyote Valley: Deep History in the High Rockies.* Cambridge, MA: Harvard University Press, 2015.

Angus, Charlie. *Cobalt: Cradle of the Demon Metals, Birth of a Mining Superpower.* Toronto: House of Anansi Press, 2022.

———. *Industrial Cathedrals of the North.* Toronto: Between the Lines, 1999.

———. *Mirrors of Stone: Fragments from the Porcupine Frontier.* Toronto: Between the Lines, 2001.

———. *We Lived a Life and Then Some.* Toronto: Between the Lines, 1996.

Armstrong, Chris. *Moose Pastures and Mergers: The Ontario Securities Commission and the Regulation of Share Markets in Canada, 1940–1980.* Toronto: University of Toronto Press, 2001.

Arnold, Jörg. "'The Death of Sympathy': Coal Mining, Workplace Hazards, and the Politics of Risk in Britain, ca. 1970–1990." *Historical Social Research* 41 (2016): 91–110.

Baeten, John. "Contamination as Artifact: Waste and the Presence of Absence at the Trout Lake Concentrator, Coleraine, Minnesota." In *Geographies of Post-industrial Place, Memory, and Heritage,* edited by Mark Alan Rhodes II, William Price, and Amy Walker, 85–102. Oxon: Routledge, 2021.

Baldwin, Douglas Owen. "Cobalt: Canada's Mining and Milling Laboratory, 1903–1918." *HSTC Bulletin: Journal of the History of Canadian Science, Technology and Medicine* 8, no. 2 (1984): 95–111.

Baldwin, Douglas Owen, and David Duke. "'A Grey Wee Town': An Environmental History of Early Silver Mining and Cobalt, Ontario." *Urban History Review* 34, no. 1 (2005): 71–87.

Banner, Stuart. *Possessing the Pacific: Land, Settlers, and Indigenous People from Australia to Alaska.* Cambridge, MA: Harvard University Press, 2007.

Barnes, Michael. *Fortunes in the Ground: Cobalt, Porcupine & Kirkland Lake.* Erin: Boston Mills Press, 1986.

———. *Gold in the Porcupine.* Cobalt, ON: Highway Bookshop, 1975.

Bateman, R., and Ontario Geological Survey. *The Timmins-Porcupine Gold Camp, Northern Ontario: The Anatomy of an Archaean Greenstone Belt and Its Gold Mineralization.* Sudbury: Ontario Geological Survey, 2005.

Beckett, Caitlynn, and Arn Keeling. "Rethinking Remediation: Mine Reclamation, Environmental Justice, and Relations of Care." *Local Environment* 24, no. 3 (2019): 216–30.

Berton, Pierre. *Klondike: The Last Great Gold Rush.* Rev. ed. Toronto: McClelland and Stewart, 1972.

Black, Megan. *The Global Interior: Mineral Frontiers and American Power.* Cambridge, MA: Harvard University Press, 2018.

Blainey, Geoffrey. *The Rush That Never Ended: A History of Australian Mining.* Parkville: Melbourne University Press, 1963.

Boag, Peter. *Re-dressing America's Frontier Past.* Berkeley: University of California Press, 2011.

Boyce, Gerald E. *Eldorado: Ontario's First Gold Rush.* Toronto: Natural Heritage/ Natural History, 1992.

Brown, L. Carson. *The Golden Porcupine.* 2nd ed. Toronto: Ontario Department of Mines, 1968.

Buckley, Karen. *Danger, Death, and Disaster in the Crowsnest Pass Mines, 1902–1928.* Calgary: University of Calgary Press, 2004.

Butler, Paula. *Colonial Extractions: Race and Canadian Mining in Contemporary Africa.* Toronto: University of Toronto Press, 2015.

Campuzano Duque, Lorena. "The People's Gold: Race and Vernacular Mining in the 'Ailing' Landscapes of Antioquia, Colombia, 1540–1958." PhD diss., Binghamton University, 2021.

Cartwright, Alan Patrick. *The Gold Miners.* Cape Town: Purnell, 1962.

Chapple, Simon. "Law and Society Across the Pacific: Nevada County, California, 1849–1860 and Gympie, Queensland, 1867–1880." PhD diss., University of New South Wales, 2010.

Coates, Ken. "The Klondike Gold Rush in World History: Putting the Stampede in Perspective." *Northern Review* 19 (Winter 1998): 21–35.

Colvine, A. C., ed. *The Geology of Gold in Ontario, Ontario Geological Survey.* Miscellaneous Paper 110. Ottawa: Ministry of Northern Development and Mines, 1983.

Cordell, Ryan, and David Smith. "Viral Texts: Mapping Networks of Reprinting in 19th-Century Newspapers and Magazines." 2017. http://viraltexts.org.

Creighton, Donald. *The Commercial Empire of the St. Lawrence, 1760–1850.* Toronto: Ryerson, 1937.

Cronon, William. *Changes in the Land: Indians, Colonists and the Ecology of New England.* New York: Hill and Wang, 2003.

Cross, Michael, ed. *The Frontier Thesis and the Canadas.* Toronto: Copp Clark, 1970.

Curtis, Kent. *Gambling on Ore: The Nature of Metal Mining in the United States.* Boulder: University of Colorado Press, 2013.

Dasmann, Raymond. "Environmental Changes Before and After the Gold rush." In *A Golden State: Mining and Economic Development in Gold Rush California,* edited by Richard Orsi and James Rawls, 105–21. Berkeley: University of California Press, 1999.

Davenport, Jade. *Digging Deep: A History of Mining in South Africa.* Cape Town: Jonathon Ball, 2013.

Davis, John. "The Klondike Gold Rush as Seen Through the British Press." *Northern Review* 19 (Winter 1998): 36–56.

Davy, Daniel Joseph. "Lost Tailings: Gold Rush Societies and Cultures in Colonial Otago, New Zealand, 1861–1911." PhD diss., University of Otago, 2013.

Demuth, Bathsheba. *Floating Coast: An Environmental History of the Bering Strait.* New York: W. W. Norton, 2019.

Drake, Karen. "The Trials and Tribulations of Ontario's 'Mining Act': The Duty to Consult and Anishnaabek Law." *McGill International Journal of Sustainable Development* 11, no. 2 (2015): 183–218.

Dunlap, Thomas. *Nature and the English Diaspora: Environment in the United States, Canada, Australia, and New Zealand.* Cambridge: Cambridge University Press, 1999.

Durand, F. J. "The Impact of Gold Mining on the Witwatersrand on the Rivers and Karst System of Gauteng and North West Province, South Africa." *Journal of African Earth Sciences* 68 (June 15, 2012): 24–43.

Fancy, Peter. *Temiskaming Treasure Trails.* Vols. 4–5. Cobalt, ON: Highway Bookshop, 1992.

Fetherling, George. *The Gold Crusades: A Social History of Gold Rushes, 1849–1929.* Toronto: University of Toronto Press, 1997.

Forestell, Nancy. "All That Glitters Is Not Gold: The Gendered Dimensions of Work, Family, and Community Life in the Northern Ontario Goldmining Town of Timmins, 1909–1950." PhD diss., University of Toronto, 1993.

———. "And I Feel Like I'm Dying from Mining for Gold: Disability, Gender, and the Mining Community, 1920–1950." *Labor: Studies in Working-Class History of the Americas* 3, no. 3 (2006): 77–93.

Fox, Julia. "Mountaintop Removal in West Virginia: An Environmental Sacrifice Zone." *Organization & Environment* 12, no. 2 (1999): 163–83.

Frost, Warwick. "The Environmental Impacts of the Victorian Gold Rushes." *Australian Economic History Review* 53, no. 1 (2013): 72–90.

Fyon, J. A., A. H. Green, and the Geological Survey of Canada. *Geology and Ore Deposits of the Timmins District, Ontario.* Field Trip 6. Ottawa: Energy, Mines, and Resources Canada, 1991.

Gabrys, Jennifer. "Sink: The Dirt of Systems." *Environment and Planning D: Society and Space* 27 (2009): 666–81.

Garrod, J. Z., and Laura Macdonald. "Rethinking 'Canadian Mining Imperialism' in Latin America." In *Mining in Latin America: Critical Approaches to the New Extraction,* edited by Kalowatie Deonanden and Michael Dougherty, 100–115. London and New York: Routledge, 2016.

Girdwood, Charles P., Lawrence Jones, and George Lonn. *The Big Dome: Over Seventy Years of Gold Mining in Canada.* Toronto: Cybergraphics, 1983.

Gomez, Rocio. *Silver Veins, Dusty Lungs: Mining, Water, and Public Health in Zacatecas, 1835–1946.* Lincoln: University of Nebraska Press, 2020.

Gordon, Todd, and Jeffrey Webber. *Blood of Extraction: Canadian Imperialism in Latin America.* Halifax: Fernwood, 2016.

———. "Canadian Capital and Secondary Imperialism in Latin America." *Canadian Foreign Policy Journal* 25, no. 1 (2019): 72–89.

Gray, John. "The Fire That Wiped Out Porcupine," *Maclean's,* February 1, 1954, 20–38.

Grossman, Sarah. *Mining the Borderlands: Industry, Capital, and the Emergency of Engineers in the Southwest Territories, 1855–1910.*. Reno: University of Nevada Press, 2018.

Groves, David, Richard Goldfarb, and Craig Hart. "Gold Deposits in Metamorphic Belts: Overview of Current Understanding, Outstanding Problems, Future Research, and Exploration Significance." *Economic Geology* 98, no. 1 (2003): 1–29.

Grunstra, Mary, and David Martell. "Along a Rickety Road: One Hundred Years of Railway Fire in Ontario's Forests." In "Forest Fire and Firefighting History." Special issue, *Forestory: Forest Fire and Firefighting History* 4, no. 1 (2013): 5–10.

Hogaboam, Dieter Grant. "Compensation and Control: Silicosis in the Ontario Hardrock Mining Industry, 1921–1975." Master's thesis, Queens University, 1997.

Hogenboom, Barbara. "Depoliticized and Repoliticized Minerals in Latin America." *Journal of Developing Societies* 28, no. 2 (2012): 133–58.

Hoogeveen, Dawn. "Sub-surface Property, Free Entry Mineral Staking and Settler Colonialism in Canada." *Antipide* 47, no. 1 (2015): 121–38.

Horowitz, Andy. *Katrina: A History, 1915–2015*. Cambridge, MA: Harvard University Press, 2020.

Innis, Harold. *The Fur Trade in Canada*. Toronto: University of Toronto Press, 1956.

Isenberg, Andrew. *Mining California: An Ecological History*. New York: Hill and Wang, 2005.

James, Marcus. "The Struggle against Silicosis in the Australian Mining Industry: The Role of the Commonwealth Government, 1920–1950." *Labor History* 65 (1993): 75–95.

Jeeves, Alan. *Migrant Labor in South Africa's Mining Economy: The Struggle for the Gold Mines' Labor Supply, 1890–1920*. Kingston: McGill–Queens University Press, 1985.

Jorgenson, Mica, and John Sandlos. "Dust Versus Dust: Aluminum Therapy and Silicosis in the Canadian and Global Mining Industries." *Canadian Historical Review* 102, no. 1 (2021): 1–26.

Katz, Elaine. "The Role of American Mining Technology and American Mining Engineers in the Witwatersrand Gold Mining Industry, 1890–1910." *South African Journal of Economic History* 20, no. 2 (2005): 48–82.

Keeling, Arn, and Patricia Boulter. "From Igloo to Mine Shaft: Inuit Labor and Memory at the Rankin Lake Nickel Mine." In *Mining and Communities in Northern Canada: History, Politics, and Memory*, edited by John Sandlos and Arn Keeling, 35–58. Calgary: Calgary University Press, 2015.

Kelley, Robert L. *Gold vs. Grain: The Hydraulic Mining Controversy in California's Sacramento Valley: A Chapter in the Decline of the Concept of Laissez Faire*. Glendale, CA: Arthur H Clark, 1959.

Klubock, Thomas. "Working-Class Masculinity, Middle-Class Morality, and Labor Politics in the Chilean Copper Mines." *Journal of Social History* 30 (1996): 435–63.

Law, John. "Notes on the Theory of the Actor-Network: Ordering, Strategy, and Heterogeneity." *Systems Practice* 5, no. 4 (1992): 379–93.

Lawrence, Susan, and Peter Davis. "The Sludge Question: The Regulation of Mine Tailings in Nineteenth-Century Victoria." *Environment and History* 20 (2014): 385–404.

LeBourdais, D. M. *Metals and Men: The Story of Canadian Mining*. Toronto: McClelland and Stewart, 1957.

LeCain, Timothy. *Mass Destruction: The Men and Giant Mines That Wired America and Scarred the Planet*. New Brunswick, NJ: Rutgers University Press: 2009.

Leech, Brian. *The City That Ate Itself: Butte, Montana, and Its Expanding Berkeley Pit*. Reno: University of Nevada Press, 2019.

Leslie, A. P. "Large Fires in Ontario Prior to 1950." In "Additional Forest Fire History Sources," edited by Sherry Hambly. Special issue, *Forestory: Forest Fire and Firefighting History* 4, no. 1 (2013): 50.

Liboiron, Max. *Pollution Is Colonialism*. Durham, NC: Duke University Press, 2021.

———. "Why Discard Studies?" *Discard Studies* (blog), July 5, 2015. https://discardstudies.com/2014/05/07/why-discard-studies/.

Long, John. *Treaty No. 9: Making the Agreement to Share the Land in Far Northern Ontario in 1905*. Montreal: McGill-Queens University Press, 2010.

Luby, Brittany. *Dammed: The Politics of Loss and Survival in Anishinaabe Territory.* Winnipeg: University of Manitoba Press, 2020.

Luxton, Mex. *More than a Labor of Love: Three Generations of Women's Work in the Home.* Toronto: Women's Press, 2009.

Macbeth, Mike. *Silver Threads among the Gold: The Rags to Riches Saga of a Man and His Mines.* Toronto: Trans-Canada Press, 1987.

MacDougall, J. B. *Two Thousand Miles of Gold, from Val d'Or to Yellowknife.* Toronto: McClelland and Stewart, 1946.

MacDowell, Laren Sefton. *An Environmental History of Canada.* Vancouver: University of British Columbia Press, 2012.

Manore, Jean L. *Cross-Currents: Hydroelectricity and the Engineering of Northern Ontario.* Waterloo, ON: Wilfred Laurier University Press, 1999.

May, Phillip Ross. *On the Motherlode.* Christchurch: University of Canterbury Press, 1971.

———. *Origins of Hydraulic Mining in California.* Oakland: Holmes Book, 1970.

———. *The West Coast Gold Rushes.* Christchurch: Pegasus Press, 1962.

McCarthy, Daniel, et al. "A First Nations–Led Social Innovation: A Moose, a Gold Mining Company, and a Policy Window." *Ecology and Society* 19, no. 4 (2014): 2.

McCullough, Jock. "Air Hunger: The 1930 Johannesburg Conference and the Politics of Silicosis." *History Workshop Journal* 72 (2011): 118–37.

———. *South Africa's Gold Mines and the Politics of Silicosis.* Rochester, NY: Boydell & Brewer, 2012.

McNeill, John, and George Vrtis, eds. *Mining North America: An Environmental History Since 1522.* Oakland: University of California Press, 2017.

Merchant, Carolyn. *Green Versus Gold: Sources in California's Environmental History.* Washington, DC: Island Press, 1998.

Milne, Courtney. *Spirit of the Land: Sacred Places in Native North America.* Toronto: Penguin Books, 1994.

Mitchell, Elaine Allan. *Fort Temiskaming and the Fur Trade.* Toronto: University of Toronto Press, 1977.

Moore, P. R., and N. A. Ritchie. "In Ground Ore-Roasting Kilns on the Hauraki Goldfield, Coromandel Peninsula, New Zealand." *Australasian Historical Archaeology* 16 (1998): 45–59.

Morrison, James. "The Robinson Treaties of 1850: A Case Study." Report prepared for the Royal Commission on Aboriginal Peoples. Ottawa: Treaty and Land Research Section, 1996.

Morse, Kathryn. *The Nature of Gold: An Environmental History of the Klondike Gold Rush.* Seattle: University of Washington Press, 2003.

Mosby, Ian. *Food Will Win the War: The Politics, Culture, and Science of Food on Canada's Home Front.* Vancouver: UBC Press, 2014.

Mouat, Jeremy. "The Genesis of Western Exceptionalism: British Columbia's Hard Rock Miners, 1895–1903." *Canadian Historical Review* 71, no. 3 (1990): 317–45.

———. "The Ultimate Crisis of the Waihi Gold Mining Company." *New Zealand Journal of History* 26, no. 2 (1992): 184–204.

Mountford, Benjamin, and Stephen Tuffnell, eds. *A Global History of Gold Rushes.* Oakland: University of California Press, 2018.

Murphy, John P. *Yankee Takeover at Cobalt!* Cobalt, ON: Highway Book Shop, 1977.

Nelles, H. V. *The Politics of Development: Forests, Mines & Hydro-electric Power, 1849–1941.* 2nd ed. Montreal: McGill–Queens University Press, 2005.

Newell, Dianne. *Technology on the Frontier: Mining in Old Ontario.* Vancouver: University of British Columbia Press, 1986.

Nixon, Rob. *Slow Violence and the Environmentalism of the Poor.* Cambridge: Harvard University Press, 2013.

Nystrom, Eric. *Seeing Underground: Maps, Models, and Mining Engineering in America.* Reno: University of Nevada Press, 2016.

Olsen, Erik. *A History of Otago.* Ann Arbour: Michigan, 1984.

Pain, S. A. *Three Miles of Gold: The Story of Kirkland Lake.* Toronto: Ryerson Press, 1960.

Paterson, John F. *Silicosis in Hardrock Miners in Ontario (a Further Study).* Bulletin 173. Toronto: Ministry of Natural Resources Ontario, 1973.

Paul, Rodman. *Mining Frontiers of the Far West, 1848–1880.* New York: Holt, Rinehart, and Winston, 1964.

Pawson, Eric, and Tom Brooking Hearn, eds. *Environmental Histories of New Zealand.* Melbourne: Oxford University Press, 2002.

Petersen, James Otto. "The Origins of Canadian Gold Mining: The Part Played by Labor in the Transition from Tool Production to Machine Production." PhD diss., University of Toronto, 1977.

Perry, Adele. *On the Edge of Empire: Gender, Race, and the Making of British Columbia, 1849–1871.* Toronto: University of Toronto Press, 2002.

Piper, Liza. *The Industrial Transformation of Subarctic Canada.* Vancouver: UBC Press, 2009.

Pyne, Stephen. *Burning Bush: A Fire History of Australia.* Seattle: University of Washington Press, 1991.

———. *Vestal Fire: An Environmental History, Told Through Fire, of Europe and Europe's Encounter with the World.* Seattle: University of Washington Press, 1997.

———. *Year of Fires: The Story of the Great Fires of 1910.* New York: Viking, 2001.

Quivik, Fredric. "Smoke and Tailings: An Environmental History of Copper Smelting Technologies in Montana, 1880–1930." PhD diss., University of Pennsylvania, 1998.

Reeves, Kier, Lionel Frost, and Charles Fahey. "Integrating the Historiography of the Nineteenth-Century Gold Rushes." *Australian Economic History Review* 50, no. 2 (2010): 111–28.

Reid, John Phillip. *Law for the Elephant: Property and Social Behaviour on the Overland Trail.* San Marino, CA: Huntington Library, 1980.

Rosner, David, and Gerald Markowitz. *Deadly Dust: Silicosis and the Politics of Occupational Disease in Twentieth-Century America.* Princeton, NJ: Princeton University Press, 1991.

Sandlos, John, and Arn Keeling. "Claiming the New North: Mining and Colonialism at the Pine Point Mine, Northwest Territories, Canada." *Environment and History* 18, no. 1 (2012): 5–34.

———, eds. *Mining and Communities in Northern Canada: History, Politics, and Memory.* Calgary: Calgary University Press, 2015.

———. "Pollution, Local Activism, and the Politics of Development in the Canadian North." *RCC Perspectives: Transformations in Environment and Society,* no. 4 (2016): 25–32.

———. "Zombie Mines and the (Over) Burden of History." *Solutions Journal* 4, no. 3 (2013). https://research.library.mun.ca/1968/.

Scott, Heidi. "Taking the Enlightenment Underground: Mining Spaces and Cartographic Representation in the Late Colonial Andes." *Journal of Latin American Geography* 14, no. 3 (2015): 7–34.

Sellers, Christopher. *Hazards of the Job: From Industrial Disease to Environmental Health Science.* Chapel Hill: University of North Carolina Press, 1997.

Smith, Donald. *Aboriginal Ontario: Historical Perspectives on the First Nations.* Toronto: Dundurn Press, 1994.

Smith, Philip. *Harvest from the Rock: A History of Mining in Ontario.* Toronto: Macmillan, 1986.

Studnicki-Gizbert, David, and David Schechter. "The Environmental Dynamics of a Colonial Fuel Rush: Silver Mining and Deforestation in New Spain, 1522 to 1810." *Environmental History* 15, no. 1 (2010): 94–119.

Surtees, Robert. "Treaty Research Report: The Robinson Treaties (1850)." Treaties and Historical Research Centre: Indian and Northern Affairs Canada, 1986. Accessed August 9, 2016. https://www.aadnc-aandc.gc.ca/eng/1100100028974/1100100028976.

Tarr, Joel. *The Search for the Ultimate Sink: Urban Pollution in Historical Perspective.* Akron: University of Akron Press, 1996.

Thompson, I. D., A. Perera, and David Euler, Ministry of Natural Resources. *Ecology of a Managed Terrestrial Landscape: Patterns and Processes of Forest Landscapes in Ontario.* Vancouver: UBC Press, 2000.

Thorpe, Jocelyn. *Temagami's Tangled Wild: Race, Gender, and the Making of Canadian Nature.* Vancouver: UBC Press, 2011.

Townsley, B. F. *Mine Finders: The History and Romance of Canadian Mineral Discoveries.* Toronto: Saturday Night Press, 1935.

Turner, Frederick Jackson. *The Significance of the Frontier in American History.* New York: Henry Holt, 1920.

Tyrell, Ian. *True Gardens of the Gods: Californian-Australian Environmental Reform, 1860–1930.* Berkeley: University of California Press, 1999.

Van Horssen, Jessica. *A Town Called Asbestos: Environmental Contamination, Health, and Resilience in a Resource Community.* Vancouver: UBC Press, 2016.

Vasiliadis, Peter. "Dangerous Truth: Interethnic Competition in a Northeastern Ontario Goldmining Center." PhD diss., Simon Fraser University, 1984.

———. *Dangerous Truth: Interethnic Competition in a Northeastern Ontario Goldmining Center.* New York: AMS Press, 1989.

Vincent, Kevin. *Bootlegged Gold: Amazing Untold Stories from the 20th Century's Lucrative World of Gold Smuggling.* Vol. 1. Self-published.

Vrtis, George. "Gold Rush Ecology: The Colorado Experience." *Journal of the West* 49, no. 2 (2010): 23–31.

Weir, Gail. *The Miners of Wabana: The Story of the Iron Ore Miners of Bell Island.* St. Johns: Breakwater Books, 1989.

White, Neil. *Company Towns: Corporate Order and Community.* Toronto: University of Toronto Press, 2012.

White, Richard. *The Organic Machine: The Remaking of the Columbia River.* New York: Hill and Wang, 1995.

Wood, David. *Places of Last Resort: The Expansion of the Farm Frontier into the Boreal Forest in Canada, c. 1910–1940.* Kingston: McGill-Queens University Press, 2006.

Ziebarth, Marilyn. "California's First Environmental Battle." *California History* 63, no. 4 (1984): 274–79.

Notes

Introduction

1. "Mountain of Gold Found by a Fluke," *Toronto Weekly Star,* December 27, 1924, 1.

2. Peter Fancy, *Temiskaming Treasure Trails,* 5:172.

3. Kent Curtis calls this complex ecology a "hybrid tangle," and Liza Piper describes it as a "blurring" between organic and inorganic labor. Kent Curtis, *Gambling on Ore: The Nature of Metal Mining in the United States,* 5; Liza Piper, *The Industrial Transformation of Subarctic Canada,* 153.

4. This argument underpins Benjamin Mountford and Stephen Tuffnell, eds., *A Global History of Gold Rushes.* Mountford and Tuffnell's collection of essays spans the period from 1858 to the turn of the twentieth century and includes Australia, Asia, North America, and Africa. However, as I will suggest, a transnationally connected mining industry arguably applies to an even wider geography and longer temporality, since it could feasibly include the mineral rushes following the colonization of Latin America as well as present explorations conducted by multinational companies on mineral frontiers around the world.

5. Kathryn Morse, *The Nature of Gold: An Environmental History of the Klondike Gold Rush,* 17.

6. Several historians have made arguments about the mobility of environmental and legal ideology between settler states. Thomas Dunlap describes the exchange of environmental ideas between British colonies and successor states. He argues that the "neo-Europes" inherited notions about nature protection from Britain. He argues that this knowledge exchange was neither a one-off nor one-way, but was part of a continuous international flow of environmental ideology largely invisible to historians. Thomas Dunlap, *Nature and the English Diaspora: Environment in the United States, Canada, Australia, and New Zealand.* Stuart Banner argues that prospectors, settlers, and governments arrived on frontiers with shared ideological baggage derived from British understandings of ownership and rights and then adapted these ideas based on local circumstances. Stuart Banner, *Possessing the Pacific: Land, Settlers, and Indigenous People from Australia to Alaska,* 5.

7. To borrow the language of actor-network theory, Porcupine and its fellow rushes became effects generated by a series of complex interactions between diverse human and nonhuman materials. John Law, "Notes on the Theory of the Actor-Network: Ordering, Strategy, and Heterogeneity," 380.

8. It is important to understand that one type of mining did not replace the other. Different types of gold deposits will always require different extractive processes, and individualistic placer-type mining was never replaced by large-scale industrial mining. Gold rushes based on placer finds continue to occur to the present (although not usually in the same places as pit mines).

9. H. V. Nelles, *The Politics of Development: Forests, Mines, and Hydro-Electric Power in Ontario, 1849–1941,* 20–21.

10. Thomas Gibson, *The Mining Laws of Ontario and the Department of Mines,* 9.

11. "Canada and the American Institute," *Engineering and Mining Journal* 115, no. 18 (1923): 787.

12. On the shape of the reefs, see Elaine Katz, "The Role of American Mining Technology and American Mining Engineers in the Witwatersrand Gold Mining Industry, 1890–1910," 61–63. On environmental and human conflict with the industry, see J. F. Durand, "The Impact of Gold Mining on the Witwatersrand on the Rivers and Karst System of Gauteng and North West Province, South Africa." On stability in the industry up to the present, see Jade Davenport, *Digging Deep: A History of Mining in South Africa,* 460–64.

13. George Fetherling, *The Gold Crusades: A Social History of Gold Rushes, 1849–1929,* 124–25; Morse, *Nature of Gold,* 138–41.

14. Timothy LeCain, *Mass Destruction: The Men and Giant Mines That Wired America and Scarred the Planet,* 20; Curtis, *Gambling on Ore,* 16.

15. "Canada Now the Mineral Treasure House to World," *Porcupine Advance,* February 16, 1928.

16. These narratives are most thoroughly developed in Australia and the United States. For Australia, see Geoffrey Blainey, *The Rush That Never Ended: A History of Australian Mining.* For the United States, see Rodman Paul, *Mining Frontiers of the Far West, 1848–1880.*

17. Turner argued that the availability of "empty" land in the American West encouraged the formation of a liberal democratic society that rejected established institutions and celebrated individualism. Frederick Jackson Turner, chap. 1 of *The Significance of the Frontier in American History.*

18. Blainey, *Rush That Never Ended,* 1–2.

19. See Michael Cross, ed., *The Frontier Thesis and the Canadas;* or D. M. LeBourdais, *Metals and Men: The Story of Canadian Mining.* Kerry Abel provides a useful analysis of frontier thesis in the context of the Porcupine/Timmins region in *Changing Places: History, Community, and Identity in Northeastern Ontario,* 400. Mining and waterpower form the foundation of Viv Nelles's explanation of the state in *The Politics of Development.*

20. Michael Barnes, *Fortunes in the Ground: Cobalt, Porcupine & Kirkland Lake;* Phillip Smith, *Harvest from the Rock: A History of Mining in Ontario.*

21. With a few important exceptions, two of the most notable being Fetherling, *Gold Crusades;* and John McNeill and George Vrtis, eds., *Mining North America: An Environmental History Since 1522.*

22. Julia Fox, "Mountaintop Removal in West Virginia: An Environmental Sacrifice Zone"; John Sandlos and Arn Keeling, "Pollution, Local Activism, and the Politics of Development in the Canadian North"; John Sandlos and Arn Keeling, "Zombie Mines and the (Over)Burden of History."

23. I am referring specifically to Jessica Van Horssen, *A Town Called Asbestos: Environmental Contamination, Health, and Resilience in a Resource Community;* and Rocio Gomez, *Silver Veins, Dusty Lungs: Mining, Water, and Public Health in Zacatecas,*

1835–1946. Other examples in the genre include Susan Lawrence and Peter Davis, "The Sludge Question: The Regulation of Mine Tailings in Nineteenth-Century Victoria"; John Sandlos and Arn Keeling, "Claiming the New North: Mining and Colonialism at the Pine Point Mine, Northwest Territories, Canada"; John Sandlos and Arn Keeling, eds., *Mining and Communities in Northern Canada: History, Politics, and Memory;* and George Vrtis, "Gold Rush Ecology: The Colorado Experience."

24. Andrew Isenberg, *Mining California: An Ecological History;* LeCain, *Mass Destruction;* Brian Leech, *The City That Ate Itself: Butte, Montana, and Its Expanding Berkeley Pit.*

25. Curtis, *Gambling on Ore;* Piper, *Industrial Transformation of Subarctic Canada.*

26. Megan Black, *The Global Interior: Mineral Frontiers and American Power;* Sarah Grossman, *Mining the Borderlands: Industry, Capital, and the Emergency of Engineers in the Southwest Territories, 1855–1910;* Eric Nystrom, *Seeing Underground: Maps, Models, and Mining Engineering in America.*

27. With a few exceptions. Transnational histories of North American mining include Thomas Andrews, *Coyote Valley: Deep History in the High Rockies;* McNeill and Vrtis, *Mining North America;* Morse, *Nature of Gold;* and Ian Tyrell, *True Gardens of the Gods: Californian-Australian Environmental Reform, 1860–1930.* Liza Piper and Bathsheba Demuth have written transnational histories of industrial extraction. Piper, *Industrial Transformation of Subarctic Canada;* Bathsheba Demuth, *Floating Coast: An Environmental History of the Bering Strait.*

28. Rob Nixon describes slow violence as the kind "that occurs gradually and out of sight, a violence of delayed destruction that is dispersed across time and space, an attritional violence that is typically not viewed as violence at all." He further points out that slow violence is exponential and tends to be underrepresented in both "strategic planning and human memory." Rob Nixon, *Slow Violence and the Environmentalism of the Poor,* 3.

29. My framing of overlapping contexts is similar to the "kaleidoscope" metaphor to Paula Butler in her work *Colonial Extractions: Race and Canadian Mining in Contemporary Africa.*

Chapter 1. "Promise of Reward to the Prospector"

1. William Arthur Parks, "Niven's Base Line, 1899," in *Report of the Bureau of Mines* (Toronto: Queen's Printer, 1900), 141.

2. As Jocelyn Thorpe argues in *Temagami's Tangled Wild: Race, Gender, and the Making of Canadian Nature,* northern Ontario possessed multiple meanings for different people at different times. However, some visions came to dominate the imaginative and physical landscape more than others, in line with social and racial structures in Canadian society.

3. As George Fetherling argues, by 1909 the Porcupine prospectors "worked against a backdrop of what by then was a familiar set of circumstances" considering the long history of gold rushes that had taken place around the world up to that point. Fetherling, *Gold Crusades,* 165.

4. Fetherling locates the rise of scientific prospecting in northern Ontario's early mineral rushes. "Generations of prospectors had been quick to disguise their past to

accept distinctions like forty-niners and sourdough, which suggested amateur status. Now the trend was reversed. Humble prospectors often made brief appearances at mining schools." Fetherling, *Gold Crusades,* 171.

5. "Metallic Minerals," Ontario Ministry of Northern Development and Mines.

6. A. C. Colvine, ed., *The Geology of Gold in Ontario, Ontario Geological Survey,* 3.

7. C. J. Hodgson, "Preliminary Report on a Computer File of Gold Deposits of the Abitibi Belt, Ontario," in *Geology of Gold in Ontario,* ed. Colvine, 32.

8. Donald Smith, *Aboriginal Ontario: Historical Perspectives on the First Nations,* 4.

9. D. Smith, *Aboriginal Ontario,* 4–15.

10. Anishnaabe place-names and origin stories suggest that they originated to the west of the future mining zone, in the Winnipeg River watershed between Lake Superior and Lake Winnipeg. Courtney Milne, *Spirit of the Land: Sacred Places in Native North America,* 22; Brittany Luby, *Dammed: The Politics of Loss and Survival in Anishinaabe Territory,* 2–6.

11. For a complete account of Anishinaabe relationships with settler resource extraction in Ontario, see Luby, *Dammed.*

12. D. Smith, *Aboriginal Ontario,* 282.

13. Elaine Allan Mitchell, *Fort Temiskaming and the Fur Trade,* 8.

14. D. Smith, *Aboriginal Ontario,* 293. This study primarily addresses settler relationships with land, so settler place-names have been retained throughout for clarity.

15. Depletion was ameliorated somewhat by conservation efforts following the Hudson's Bay Company monopoly in 1821. D. Smith, *Aboriginal Ontario,* 302, 307, 317.

16. John Long, *Treaty No. 9: Making the Agreement to Share the Land in Far Northern Ontario in 1905,* 3, 36, 54.

17. Long, *Treaty No. 9,* 11, 84, 454n3.

18. Goldcorp, the operator working in Timmins/Porcupine as of writing, has formal agreements dating to 2014 with Mattagami, Wahgoshig, Matachewan, and Flying Post First Nations. Goldcorp, "Responsible Mining"; Goldcorp, "Goldcorp Signs Important Resource Development Agreement with Four First Nation Communities in Timmins, Ontario," news release, November 24, 2014.

19. Robert Surtees, "Treaty Research Report: The Robinson Treaties (1850)." This was the first time a treaty had occurred in response to resource development as opposed to settler desires for homesteads. James Morrison, "The Robinson Treaties of 1850: A Case Study," 7.

20. Surtees, "The Robinson Treaties (1850)."

21. Long, *Treaty No. 9,* 27–28, 67, 83.

22. Karen Drake, "The Trials and Tribulations of Ontario's 'Mining Act': The Duty to Consult and Anishinaabek Law"; Hereward Longley, "Indigenous Battles for Environmental Protection and Economic Benefits during the Commercialization of the Alberta Oil Sands, 1967–1986," in *Mining and Communities in Northern Canada,* ed. Sandlos and Keeling, 207–32; Daniel D. P. McCarthy et al., "A First Nations–Led Social Innovation: A Moose, a Gold Mining Company, and a Policy Window."

23. The Mining Act of 1891 created the Ontario Bureau of Mines "to aid in promoting the mining interests of the province." It worked under the Department of Crown Lands under director Archibald Blue. The Geological Branch of the new bureau engaged in summer fieldwork "in any areas which it may be thought advisable [by the

provincial geologist] to explore" and was explicitly "for the benefit of the public." The area east of the Mattagami River and North of Niven's base line first fell into this category in 1899. An exploration party usually included a head (Parks in 1899) and three assistants chosen from university students in geology or mining engineering, paid and provisioned by the bureau. Gibson, *Mining Laws of Ontario and the Department of Mines,* 111.

24. He returned in 1900 to add further details.

25. It is impossible to know if these were natural or human caused, but were most likely a combination of both. William Arthur Parks, "The Nipissing-Algoma Boundary," in *Ontario Bureau of Mines,* vol. 8, pt. 1 (Toronto: Warwick Bro's and Rutter, 1899), 176, 177, 179, 181.

26. Parks, "The Nipissing-Algoma Boundary," 178, 179, 181.

27. Kay, "The Abitibi Region," in *Thirteenth Report of the Ontario Bureau of Mines, 1904* (Toronto: King's Printer, 1904), 105, 107, 128.

28. "Larch Sawfly," Government of Canada, last modified 4 August 2015, https://tidcf.nrcan.gc.ca/en/insects/factsheet/7907.

29. Kay, "The Abitibi Region," 105, 127.

30. Kay, "The Abitibi Region," 107, 110.

31. James McMillan, "Explorations in Abitibi," in *Report of the Bureau of Mines, 1905,* vol. 14, pt. 1 (Toronto: L. K. Cameron, 1905), 185.

32. McMillan, "Explorations in Abitibi," 189.

33. Parks, "The Nipissing-Algoma Boundary," 193.

34. Kay, "The Abitibi Region," 105, 127.

35. McMillan, "Explorations in the Abitibi," 192.

36. McMillan, "Explorations in the Abitibi," 239, 243.

37. H. L. Kerr, "Exploration in Mattagami Valley," in *Report of the Bureau of Mines,* vol. 15, pt. 1 (Toronto: L. K. Cameron, King's Printer, 1906), 131, 151.

38. David Wood, *Places of Last Resort: The Expansion of the Farm Frontier into the Boreal Forest in Canada, c. 1910–1940,* 10, 13.

39. Nelles, *Politics of Development,* 118.

40. Wood, *Places of Last Resort,* 35; Nelles, *Politics of Development,* 53.

41. Parks, "The Nipissing-Algoma Boundary," 193; McMillan, "Explorations in Abitibi," 243.

42. Kerr, "Explorations in Mattagami Valley," 27, 135.

43. McMillan, "Explorations in Abitibi," 138.

44. Kerr, "Exploration in Mattagami Valley," 27, 135.

45. Parks, "Niven's Base Line, 1899," 142.

46. Kerr, "Exploration in Mattagami Valley," 127, 128.

47. Kerr wrote: "Peat may be prepared for the market at such a figure as to have a slight advantage over coal." Kerr, "Exploration in Mattagami Valley," 128.

48. Kerr, "Exploration in Mattagami Valley," 128–29.

49. Noah Timmins, for example, used money raised at Cobalt to start his mine at Porcupine. Fancy, *Temiskaming Treasure Trails,* 5:1.

50. Willet Miller, "The Cobalt-Nickel Arsenides and Silver Deposits of Temiskaming (Cobalt and Adjacent Areas)," fourth addition in *Report of the Bureau of Mines,* vol. 19, pt. 2 (Toronto: Kings Printers, 1913), xi.

51. *Nineteenth Annual Report of the Bureau of Mines, 1910,* vol. 19, pt. 1 (Toronto: L. K. Cameron, Kings Printer, 1910), 9.

52. Barnes, *Fortunes in the Ground,* 15.

53. Grubstake refers to money prospectors used to buy food, lodging, and equipment for exploring. Barnes, *Fortunes in the Ground,* 15–16.

54. John P. Murphy, *Yankee Takeover at Cobalt!,* 6.

55. Barnes, *Fortunes in the Ground,* 17.

56. Barnes, *Fortunes in the Ground,* 32.

57. "New Eldorado in a Northern Ontario Wilderness: How the Mishap of a Burly Blacksmith Revealed the Wonderful Mineral Wealth of the Cobalt District," *New York Times,* 27 May 1906.

58. "Otago School of Mines Annual Report," *Otago Daily Times,* 14 May 1910, 5.

59. Douglas Owen Baldwin, "Cobalt: Canada's Mining and Milling Laboratory, 1903–1918," 95–96, 101–2.

60. Barnes, *Fortunes in the Ground,* 28; Douglas Baldwin and David Duke, "'A Grey Wee Town': An Environmental History of Early Silver Mining and Cobalt, Ontario," 72.

61. Barnes, *Fortunes in the Ground,* 36. Descriptions of the extensive environmental damage that occurred at Cobalt can also be found in Baldwin and Duke, "'Grey Wee Town'"; Barnes *Fortunes in the Ground,* 36–63; and Mike Macbeth, *Silver Threads Among the Gold: The Rags to Riches Saga of a Man and His Mines,* 17, 18. On power production at Cobalt, see Baldwin, "Cobalt," 98–100.

62. "Cobalt: Where Streets Are Lined with Silver," *Star,* March 19, 1910, 1.

63. Daniel Joseph Davy, "Lost Tailings: Gold Rush Societies and Cultures in Colonial Otago, New Zealand, 1861–1911"; Jeremy Mouat, "The Ultimate Crisis of the Waihi Gold Mining Company"; P. R. Moore and N. A. Ritchie, "In Ground Ore-Roasting Kilns on the Hauraki Goldfield, Coromandel Peninsula, New Zealand."

64. As modern surveys of the Porcupine region have concluded, "almost all of the major producing and past-producing deposits from which most of the gold ore in the area was obtained were found by surface prospecting in the first 10 years of major exploration (1905–1914)." Nonvisible indicators would not shape prospector behavior until much later in the twentieth century and the development of aerial prospecting with powerful radar. C. J. Hodgson, "Preliminary Report on a Computer File of Gold Deposits of the Abitibi Belt, Ontario," in *Geology of Gold in Ontario,* ed. Colvine, 11.

65. "Special Report on the Porcupine," in *Twentieth Annual Report of the Ontario Bureau of Mines, 1911,* vol. 20, pt. 2 (Toronto: King's Printer, 1911), 22.

66. Free entry still underpins mining in most of North America, although recent court challenges in British Columbia and the Yukon now suggest that Indigenous consent is required prior to staking. "B.C. Fails to Meet Indigenous Consent Standard for Mining—8 Recent Cases," Mining Watch Canada, December 6, 2021, https://miningwatch.ca/news/2021/12/6/bc-fails-meet-Indigenous-consent-standard-mining-8-recent-cases.

67. Dawn Hoogeveen, "Sub-surface Property, Free Entry Mineral Staking and Settler Colonialism in Canada."

68. *Eighteenth Annual Report of the Bureau of Mines, 1909,* vol. 18, pt. 1 (Toronto: L. K. Cameron, King's Printer, 1909), 52–53.

69. Barnes, *Fortunes in the Ground,* 87, 88.

70. Barnes, *Fortunes in the Ground*, 89.

71. Given Wilson's dependence on Tom Fox's assistance, other prospectors may similarly have depended on Indigenous people to access the Porcupine in 1909. "Tom Fox—Hired by First Porcupine Prospecting Group," *Porcupine Advance*, 13 May 1926.

72. Ironically, the Hollinger claims were right on top of the lapsed D'Aigle claims. Barnes, *Fortunes in the Ground*, 92.

73. Barnes, *Fortunes in the Ground*, 93.

74. Dome, Hollinger, and McIntyre became "the big three" that dominated Porcupine mining. These were not the only claims staked at Porcupine in 1909, but other prospectors did not experience the same fantastic successes as those associated with the Dome, Hollinger, and McIntyre mines, and thus have not been well recorded by documentary evidence.

75. Nelles, *Politics of Development*, 20–23, 30.

76. *Twentieth Annual Report of the Bureau of Mines, 1911*, 271, 270.

77. *Nineteenth Annual Report of the Bureau of Mines, 1910*, 92.

78. "Toronto," *Engineering and Mining Journal*, October 16, 1909, 797.

79. "Toronto," *Engineering and Mining Journal*, November 27, 1909, 1087.

80. Fancy, *Temiskaming Treasure Trails*, 4:172, 173.

81. "The Stampede to Porcupine Gold Fields," *Northern Miner*, April 29, 1910.

82. Harold Palmer Davis, *The Davis Handbook of the Porcupine Gold District*, 17.

83. *Nineteenth Annual Report of the Bureau of Mines, 1910*, 11.

84. *Nineteenth Annual Report of the Bureau of Mines, 1910*, 11, 12.

85. Thomas Gibson, "Ontario," *Engineering and Mining Journal*, January 8, 1910, 125.

86. "Porcupine Lake Region, Ontario," *Engineering and Mining Journal* (January 22, 1910): 210.

87. According to Barnes, "Cyril Knight made the first map of the Hollinger property and at that time found it difficult to draw boundaries between the intrusive porphyry and the basic lavas." Barnes, *Fortunes in the Ground*, 99.

88. *Twentieth Report of the Bureau of Mines, 1911*, 260.

89. A. G. Burrows, "The Porcupine Gold Area," in *Report of the Bureau of Mines, 1911* (Part II) (Toronto: King's Printer, 1911), 20.

90. "Alex Kelso Discoverer of Alexo Nikel Mine and Pioneer of the North," n.d., file 3, MG 30 E 219, Earnest F. Pullen Fonds, LAC (hereafter cited as LAC), 3.

91. "Fox, Tom—Hired by First Porcupine Prospecting Group," *Porcupine Advance*, 13 May 1926.

92. "Mountain of Gold Found by Fluke," *Toronto Weekly Star*, December 27, 1924, 1.

93. Jack Campbell, "The Treasure Chests of Canada Located in Northern Ontario, Found by Greenhorn Prospectors and Financed by Greenhorn Businessmen," n.d., file 1-22, Campbell, Jack, Letter Describing Discovery of Porcupine Mining Area, vol. 1, MG 30 W. S. Edwards Fonds, LAC.

94. "Fox, Tom—Hired by First Porcupine Prospecting Group."

95. He did briefly describe local possibilities for water power. Burrows, "The Porcupine Gold Area," 32–33.

96. On the maps, see *Nineteenth Annual Report of the Bureau of Mines, 1910*, 12. On the new towns, see *Nineteenth Annual Report of the Bureau of Mines, 1910*, 120.

97. *Nineteenth Annual Report of the Bureau of Mines, 1910*, 9.

98. *Engineering and Mining Journal,* November 13, 1909, 995.

99. Barnes, *Fortunes in the Ground,* 28.

100. Barnes, *Fortunes in the Ground,* 92.

101. Barnes, *Fortunes in the Ground,* 101.

102. *Nineteenth Annual Report of the Bureau of Mines, 1910,* 120.

103. Barnes, *Fortunes in the Ground,* 99; James Otto Petersen, "The Origins of Canadian Gold Mining: The Part Played by Labor in the Transition from Tool Production to Machine Production," 150.

104. C. W. Knight, "The Destruction of Valuable Water Power on the Frederick House River, near Porcupine," *Canadian Mining Journal* (February 1, 1911): 91–92.

105. Fancy, *Temiskaming Treasure Trails,* 5:39.

106. Knight, "Destruction of Valuable Water Power," 91–92.

107. "Porcupine Lake Region, Ontario," 209.

108. Fancy, *Temiskaming Treasure Trails,* 5:1.

109. "Cobalt," *Engineering and Mining Journal* (December 25, 1909): 1289; Fancy, *Temiskaming Treasure Trails,* 5:2. In the nineteenth century, the government of Ontario had purchased diamond drills, which they lent out to small mining operations for exploration. By 1910 this program had clearly become defunct—not only did most of the companies engaged in exploration have enough capital to purchase their own diamond drills, but many private drilling companies could easily be hired to do drilling work. *Nineteenth Annual Report of the Bureau of Mines, 1910,* 25.

110. "Porcupine Lake Region, Ontario," 209.

111. Fancy, *Temiskaming Treasure Trails,* 5:3; *Nineteenth Annual Report of the Ontario Bureau of Mines, 1910,* 50.

112. "Cobalt," *Engineering and Mining Journal,* February 19, 1910, 435.

113. "Toronto," *Engineering and Mining Journal,* February 12, 1910, 387.

114. "Six of the claims were secured by the Dome Mines Company, of New York and Copper Cliff, for $1,500,000, the remaining five being taken by McCormick Brothers, of New York, at $200,000." "Toronto," *Engineering and Mining Journal,* April 23, 1910, 886.

115. "Toronto," *Engineering and Mining Journal,* February 19, 1910, 434.

116. According to the *Journal,* "Free gold was recently found by John Callahan, 3½ miles east of Night Hawk lake, the vein being reported to be 8 ft. wide with a fair showing of gold. This is the first find east of Night Hawk Lake." "Toronto," *Engineering and Mining Journal,* February 19, 1910, 435.

117. "Cobalt," *Engineering and Mining Journal,* February 19, 1910, 434; Fancy, *Temiskaming Treasure Trails,* 5:18. However, the *Engineering and Mining Journal* seemed confident about the project much earlier. "The government will build a branch line into Whitney and Tisdale township." "Toronto," *Engineering and Mining Journal,* March 12, 1910, 584.

118. *Nineteenth Annual Report of the Bureau of Mines, 1910,* 92; "The Porcupine District: Special Correspondence," *Engineering and Mining Journal,* April 23, 1910, 874.

119. "Toronto," *Engineering and Mining Journal,* May 5, 1910, 934.

120. "The Porcupine District: Special Correspondence," *Engineering and Mining Journal,* April 23, 1910, 874.

121. "Special Report on the Porcupine," 3.

122. "Porcupine Lake Region, Ontario," 210.

123. "Toronto," *Engineering and Mining Journal*, 5 May 1910, 934.

124. "Real Canada—Uninviting Place," *Darling Downs Gazette*, 12 May 1911, http://nla.gov.au/nla.news-article182678881.

125. "Real Canada—Uninviting Place."

126. For a similar account of the individual disappointments in the Klondike gold rush, see Ken Coates, "The Klondike Gold Rush in World History: Putting the Stampede in Perspective." In the spring of 1911, the annual report for the Bureau of Mines noted that "mining companies in the initial stage of development work on their properties are often more negligent than after the mines have begun production" when it came to worker safety. *Twentieth Annual Report of the Ontario Bureau of Mines, 1911*, 61.

127. Morse, *Nature of Gold*, 130–37.

128. Piper, *Industrial Transformation of Subarctic Canada*, 122.

129. For data and discussion on how stories crossed borders in the nineteenth century, see Ryan Cordell and David Smith, "Viral Texts: Mapping Networks of Reprinting in 19th-Century Newspapers and Magazines."

130. "Our Canadian Letter," *Otago Daily Times*, December 24, 1909, 8, http://paperspast.natlib.govt.nz/newspapers/ODT19091224.2.75; "Gold Rush Now On: All about the 'New Klondyke' Which Prospectors Have Discovered Near Ontario," *Woodville Examiner*, 3 June 1910, 5, http://paperspast.natlib.govt.nz/newspapers/WOODEX19100603.2.27.4; "Chances of Finding a Nugget," *Goulburn Evening Penny Post*, 29 July 1911, http://nla.gov.au/nla.news-article137042898; "Curb Hits a Stock Boom," *New York Times*, April 1, 1911, 22; "Porcupine Gold Camp," *Northern Miner*, June 2, 1911, http://nla.gov.au/nla.news-article80353450.

131. "Cobalt: Where Streets Are Lined with Silver"; "Mining," *Sun*, March 5, 1911, 6 http://nla.gov.au/nla.news-article211096643.

132. "The Gold Rush in the North," *Globe*, October 28, 1909, 5.

133. "Cobalt Engineers after Impostors," *Globe*, December 4, 1909, 11.

134. "The Gold Rush in the North."

135. "The New Gold Fields," *Globe*, November 3, 1909, 1.

136. "New Goldfield, Many Free Gold Veins," *Press*, January 16, 1911, 7, http://paperspast.natlib.govt.nz/newspapers/CHP19110116.2.27.33; "Display Ad 53," *New York Times*, March 12, 1911, sec. "Real Estate Financial Business News"; "Wanted—Investors for Good Cobalt Silver and Porcupine Gold," *San Francisco Call*, April 3, 1910, 54.

137. "Gold Rush Now On."

138. "General Cables," *Hawera & Normandy Star*, November 17, 1910, 7, http://paperspast.natlib.govt.nz/newspapers/HNS19101117.2.57; "Hardships of Prospectors: Starvation in Northern Ontario," *Colonist*, November 18, 1910, 2, http://paperspast.natlib.govt.nz/newspapers/TC19101118.2.25; "Gold Rush Death of Prospectors," *Akaroa Mail and Banks Peninsula Advisor*, November 18, 1910, 2, http://paperspast.natlib.govt.nz/newspapers/AMBPA19101118.2.9.10; "Prospectors' Terrible Fate: Supplies Run Out. Scores Dead and Dying," *Wanganui Herald*, November 18, 1910, http://paperspast.natlib.govt.nz/newspapers/WH19101118.2.29; "The Lure of Gold: Deaths on the Trail," *New Zealand Herald*, November 18, 1910, 5, http://paperspast.natlib.govt.nz/newspapers/NZH19101118.2.51; "The Thirst for Gold," *Grey River Argus*, November 18, 1910, 5, http://paperspast.natlib.govt.nz/newspapers/GRA19101118.2.24.7; "Trail

of Death: Unfortunate Prospectors, Shortage of Supplies," *Otago Daily Times,* November 18, 1910, 5, http://paperspast.natlib.govt.nz/newspapers/ODT19101118.2.39.

139. "Awful Suffering: Early California Again. Twenty Dead," *Daily Herald,* November 18, 1910, http://nla.gov.au/nla.news-page10493134. The same story ran in the *North West Post,* November 19, 1910, under the title "Canada Porcupine Gold Camp. Twenty Miners Dead. Sufferings Terrible," http://nla.gov.au/nla.news-article202034309.

140. Including in Australia "Canadian Winter. Tragic Story from Ontario," *Bendigo Advisor,* November 18, 1910, http://nla.gov.au/nla.news-article90686829; "Lust for Gold," *Riverine Herald,* November 18, 1910, http://nla.gov.au/nla.news-article114915843; "Mortality Among Gold Prospectors," *Mainland Daily Mercury,* November 17, 1910, retrieved from National Archives of Australia's Trove, http://nla.gov.au/nla.news-article124268728.

141. The *Clarence and Richmond Examiner* reported on the nineteenth of November that "scores of prospectors are lying dead, and many other dying of starvation through exposure on the trails to the South Porcupine gold camps in Alaska, in aid for which appeals were made but failed to come." "In the Rush for Gold," *Clarence and Richmond Examiner,* November 19, 1910, http://nla.gov.au/nla.news-article61531847.

142. "The Quest for Gold," *Dominion,* November 19, 1910, 5, http://paperspast.natlib.govt.nz/newspapers/DOM19101119.2.39; "The Porcupine Gold Rush," *Hawera & Normandy Star,* November 19, 1910, 5, http://paperspast.natlib.govt.nz/newspapers/HNS19101119.2.31.2.

143. "Engulfed in Mud. Gold Seekers Sink in Presence of Onlookers," *World's News,* January 21, 1911, http://nla.gov.au/nla.news-article128264977.

144. John Davis, "The Klondike Gold Rush as Seen Through the British Press."

145. *Twentieth Annual Report of the Bureau of Mines, 1911,* 3.

146. H. Davis, *Davis Handbook to the Porcupine Mines,* 24–26.

Chapter 2. The Great Fire

1. A. P. Leslie, "Large Fires in Ontario Prior to 1950."

2. "Porcupine Disaster Intensifies; Refugees Fleeing from the Scene," *Globe,* July 14, 1911.

3. John Gray, "The Fire That Wiped Out Porcupine."

4. I. D. Thompson, A. Perera, and David Euler, Ministry of Natural Resources, *Ecology of a Managed Terrestrial Landscape: Patterns and Processes of Forest Landscapes in Ontario,* 41–42.

5. Thompson, Perera, and Euler, *Ecology of a Managed Terrestrial Landscape,* 116.

6. Tennyson Jarvis, "Agricultural Capabilities of the Abitibi," in *Report of the Bureau of Mines, 1904* (Toronto: King's Printer, 1904), 122, 128.

7. Possibly because they visited at different times of year, or possibly because the attack was over. James McMillan, "Explorations in Abitibi," in *Report of the Bureau of Mines Annual Report, 1905* (Toronto: King's Printer, 1905), 189; A. G. Burrows, "The Porcupine Gold Area," in *Twentieth Annual Report of the Bureau of Mines, 1911,* pt. 2 (Toronto: King's Printer, 1911), 5; A. G. Burrows, "The Porcupine Gold Area Second Report," in *Ontario Bureau of Mines Annual Report, 1912* (Toronto: King's Printer, 1912), 208.

8. Stephen Pyne, *Year of Fires: The Story of the Great Fires of 1910.*

9. William Arthur Parks, "The Nipissing-Algoma Boundary," in *Ontario Bureau of Mines, 1899* (Toronto: Warwick Bro's & Rutter, 1988), 178, 182, 183.

10. A. P. Coleman, "The Northern Nickel Range," in *Ontario Bureau of Mines Annual Report, 1904* (Toronto: King's Printer, 1904), 221.

11. Mary Grunstra and David Martell, "Along a Rickety Road: One Hundred Years of Railway Fire in Ontario's Forests"; Thompson, Perera, and Euler, *Ecology of a Managed Terrestrial Landscape*, 242.

12. Thompson, Perera, and Euler, *Ecology of a Managed Terrestrial Landscape*, 120.

13. "Fire Rangers, Work Of," *Porcupine Advance*, June 25, 1915.

14. Railways are known to increase fire risk, a fact well understood by governments and citizens in the early twentieth century. See, for example, "Protecting Forests: Regulations Recommended to Govern Railways," *Globe*, May 3, 1910; "Railways and Forest Fires," *Globe*, June 6, 1904; "Railways Responsible: Locomotives Start 30% of Forest Fires," *Globe*, December 29, 1910.

15. Gray, "Fire That Wiped Out Porcupine."

16. Leslie, "Large Fires in Ontario Prior to 1950," 50.

17. George Kay, "The Abitibi Region," in *Report of the Bureau of Mines, 1904* (Toronto: King's Printer, 1904), 109, 115; James McMillan, "Explorations in Abitibi," in *Report of the Bureau of Mines, 1905* (Toronto: King's Printer, 1905), 193; H. L. Kerr, "Exploration in Mattagami Valley," in *Report of the Bureau of Mines, 1906* (Toronto: King's Printer, 1906), 124; Archibald Henderson, "Agricultural Resources of Mattagami," in *Report of the Bureau of Mines, 1906* (Toronto: King's Printer, 1906), 140.

18. J. M. Bell, "Iron Ranges of Michipicoten West," in *Report of the Bureau of Mines, 1905* (Toronto: King's Printer, 1905), 296.

19. W. L. Goodwin, "Summer Mining Classes," in *Report of the Bureau of Mines, 1907* (Toronto: King's Printer, 1907), 51, 52.

20. Stephen Pyne, *Vestal Fire: An Environmental History, Told Through Fire, of Europe and Europe's Encounter with the World*, 332.

21. Stephen Pyne, *Burning Bush: A Fire History of Australia*, 294.

22. See Pyne, *Year of Fires*.

23. "Forest Fires in America," *Gympie Times and Mary River Mining Gazette*, July 15, 1911.

24. Pyne, *Year of Fires*.

25. Government of Canada, "Ottawa Data."

26. The dates March 31, 1910, to March 30, 1911, recorded 27.72 inches total precipitation, and March 31, 1911, to March 30, 1912, recorded 29.95 inches. The yearly average 1890 to 1912 was 34.84 inches. J. H. Grisdale, "Report of the Director," in *Annual Report of the Experimental Farms for the Year Ending March 31, 1912* (Ottawa: King's Printer, 1912), 17.

27. "Record Is Now Fifty-Eight Deaths," *Globe*, July 8, 1911.

28. W. T. Macoun, "Report of the Dominion Horticulturalist," 86.

29. Burrows, "The Porcupine Gold Area" (1911), 5.

30. Hollinger Gold Mines Limited First Annual Report, January 1912, F113 Hollinger Company Fonds, Archives of Ontario.

31. Michael Barnes, *Gold in the Porcupine*, 38.

32. See "Pail Brigade, Porcupine Fire," CPC-00731, and "Fighting Fire Last Stand Powells Point," CPC-00712, William Ready Division of Archives and Research Collections.

33. Barnes, *Gold in the Porcupine*, 38.

34. "Telegraph for Charles Murphy, Ottawa," July 27, 1911, vol. 150, file 1703, Secretary of State Correspondence 1911, LAC (hereafter cited as LAC).

35. For example, "Dome Gold Mine Plant After Porcupine Fire," C-312-0-0-0-11, C 321 H. Peters Fonds, LAC.

36. J. B. MacDougall, *Two Thousand Miles of Gold, from Val d'Or to Yellowknife*, 181.

37. "Outside Securities," *New York Times*, July 14, 1911.

38. "Pictures Horrors of Forest Fire," *New York Times*, July 16, 1911.

39. "In Fire-Girt Lake with Forest Beasts," *New York Times*, July 18, 1911.

40. "Fire Losses at Porcupine," *Wall Street Journal*, July 18, 1911.

41. *Twenty-First Annual Report of the Bureau of Mines, 1912*, 153.

42. For example, "Worst in Canada's History," *Maitland Daily Mercury*, July 14, 1911, retrieved from the National Archives of Australia's Trove, http://nla.gov.au/nla.news-article121470933; "Cheat Forest Fires," *Age*, July 15, 1911, retrieved from the National Archives of Australia's Trove, http://nla.gov.au/nla.news-article196207743.

43. "Cheat Forest Fires," *Age*, July 15, 1911.

44. "A Canadian Calamity," *Armidale Express and New England General Advertiser*, July 18, 1911, retrieved from the National Archives of Australia's Trove, http://nla.gov.au/nla.news-article191946484.

45. "The Forest Fires," *Daily Post*, July 17, 1911, retrieved from the National Archives of Australia's Trove, http://nla.gov.au/nla.news-article178345671.

46. "The Forest Fires," *Daily Advertiser*, July 15, 1911, retrieved from the National Archives of Australia's Trove, http://nla.gov.au/nla.news-article142491285.

47. "Forest Fires," *Daily Herald*, July 17, 1911, retrieved from the National Archive of Australia's Trove, http://nla.gov.au/nla.news-article105323367.

48. "Forests Become Furnaces," *Geelong Advertiser*, July 15, 1911, retrieved from the National Archives of Australia's Trove, http://nla.gov.au/nla.news-article150094783; "Forest Fires in America," *Gympie Times and Mary River Mining Gazette*, July 15, 1911, retrieved from the National Archives of Australia's Trove, http://nla.gov.au/nla.news-article190878609.

49. "Forest Fires," *Daily Post*, July 19, 1911, retrieved from the National Archives of Australia's Trove, http://nla.gov.au/nla.news-article178346141.

50. "Forest Fires," *Daily Post*, July 19, 1911.

51. "500 Burned. Worst Fire Ever Known in Canada," *Gungandai Times and Tumut, Adelong and Murrumbidgee District Advisor*, September 29, 1911, retrieved from National Archives of Australia's Trove, http://nla.gov.au/nla.news-page12725090.

52. "Warning to Sportsmen," *Victoria Daily Colonist*, September 7, 1911.

53. Canada, Statistics Canada, 1911 Census, Ottawa, 1911. Note that in 1911, origin indicated "racial or tribal origin" (traced through the father) rather than citizenship, which was listed separately.

54. Enquiries Regarding Porcupine Fire, RG 6, vol. 150, file 1703, Secretary of State Correspondence 1911, LAC.

55. "The First Wedding in Porcupine," Porcupine, Ontario, Canada, CPC-03090, William Ready Division of Archives and Research Collections.

56. "Stock Exchange—Porcupine—A Has Been," October 4, 1910, Porcupine, Ontario, Canada, CPC-00729, William Ready Division of Archives and Research Collections.

57. "Blacksmith Shop, Timons [sic] Gold Mine," CPC-00708, William Ready Division of Archives and William Ready Division of Archives and Research Collections.

58. "Mary Coutts to Secretary of State, Ottawa," July 14, 1911, RG 6, vol. 150, file 1703, Secretary of State Correspondence 1911, LAC. Three weeks later, Wilde was reported safe by the Porcupine relief committee. "Relief Committee to Thomas Mulvey," August 1, 1911, RG 6, vol. 150, file 1703, Secretary of State Correspondence 1911, LAC.

59. "P. T. Bolan to Secretary of State," July 16, 1911, RG 6, vol. 150, file 1703, Secretary of State Correspondence 1911, LAC.

60. Their skepticism was understandable considering the frequent errors in the existing reports. The Porcupine Relief Committee, formed in the days after the fire, seems to have responded to requests from Mulvey on an ad hoc basis. Their telegrams out were confusing, full of misspellings, and often gave inconsistent amounts of detail.

61. "K. V. Taylor to Lord Strathcona," July 17, 1911, RG 6, vol. 150, file 1703, Secretary of State Correspondence 1911, LAC.

62. The comma in the death list had been an error. "Thomas Mulvey, Secretary of State, to Mr. Stanley H. Moon," August 1, 1911, RG 6, vol. 150, file 1703, Secretary of State Correspondence 1911, LAC.

63. "Secretary of State to Relief Committee," September 11, 1911, RG 6, vol. 150, file 1703, Secretary of State Correspondence 1911, LAC.

64. Charlie Angus, *Cobalt: Cradle of the Demon Metals, Birth of a Mining Superpower;* Peter Boag, *Re-dressing America's Frontier Past;* Adele Perry, *On the Edge of Empire: Gender, Race, and the Making of British Columbia, 1849–1871;* Mex Luxton, *More than a Labor of Love: Three Generations of Women's Work in the Home.*

65. "Murphy to Dominion, London, England," July 14, 1911, RG 6, vol. 150, file 1703, Secretary of State Correspondence 1911, LAC.

66. "S. White, Mayor to Hon. Chas Murphy," July 13, 1911, RG 6, vol. 150, file 1703, Secretary of State Correspondence 1911, LAC; "Telegraph for Charles Murphy, Ottawa," July 27, 1911, RG 6, vol. 150, file 1703, Secretary of State Correspondence 1911, LAC.

67. As the men profiled by the *New York Times* had. "Pictures Horrors of Forest Fire," *New York Times*, July 16, 1911; "In Fire-Girt Lake with Forest Beasts."

68. Burrows, "The Porcupine Gold Area," 95–96.

69. Burrows, "The Porcupine Gold Area," 101.

70. Harkness to Edwards, "Re. Point Claims," August 22, 1911, file 2-23 Correspondence, General Business, MG 30 William S. Edwards Fonds, LAC.

71. "Greatest Mining Men Are Now in Porcupine," *Porcupine Press*, August 5, 1911, file 2-9 Newspaper Clippings, 1911–24, MG 30 William S. Edwards Fonds, LAC.

72. "Greatest Mining Men."

73. Harkness to Edwards, "Re. Whitney Claims."

74. Discussed in detail below. For example, Pearl Lake mining company (bought by McIntyre in 1912). "Casualties," *Canadian Mining Journal* 33 (May 1, 1912): 298. And

Jupiter, also bought by McIntyre in 1915. "McIntyre and Jupiter," *Canadian Mining Journal* 36 (November 1, 1915): 668.

75. Harkness to Edwards, "Re. Whitney Claims"; Weiss to Edwards, August 28, 1911, 2-23 Correspondence, General Business, MG 30 William S. Edwards Fonds, LAC; Allan to Edwards, October 8 and 24, 1911, file 2-23 Correspondence, General Business, MG 30 William S. Edwards Fonds, LAC.

76. Allan to Edwards, October 24, 1911.

77. A process likely facilitated by their foreign management, ownership, and financing.

78. W. S. Edwards, *First Report Dome Mines Company Limited, 1912*, 1–32 Dome Mines Company, LTD—First Report, MG 30 William S. Edwards Fonds, LAC.

79. James Otto Petersen includes a lengthy discussion of these men's influence on the decision to adopt tube and ball mills at Porcupine. Petersen, "Origins of Canadian Gold Mining," 51, 183–85, 217, 379–406; Dianne Newell, *Technology on the Frontier: Mining in Old Ontario*, chap. 3.

80. Edwards, *First Report Dome Mines Company Limited, 1912.*

81. *Hollinger Gold Mines Limited First Annual Report*, January 1911, F113 Hollinger Company Fonds, Archives of Ontario, 14.

82. *Hollinger Gold Mines Limited First Annual Report*, January 1913, F113 Hollinger Company Fonds, Archives of Ontario, 22.

83. For other examples, see Timothy LeCain's discussion of the spread Daniel Jackling's glory hole in *Mass Destruction.* See also Nystrom, *Seeing Underground.*

84. Edwards, *First Report Dome Mines Company Limited, 1912.*

85. *Hollinger Gold Mines Limited First Annual Report*, January 1912, F113 Hollinger Company Fonds, Archives of Ontario, 20.

86. Harkness to Edwards, August 28, 1911, file 2-23 Correspondence, General Business, MG 30 William S. Edwards Fonds, LAC; Allan to Edwards, October 8 and 15, 1911, file 2-23 Correspondence, General Business, MG 30 William S. Edwards Fonds, LAC.

87. "Shumacher Mine Filled with Water," *Porcupine Advance*, September 6, 1912.

88. H. C. Peters, "A House Afloat, South Porcupine Flood," C 312-0-0-0-3; "South Porcupine Station—Flood Time," C 312-0-0-0-45; "S. Porcupine Flood," C 312-0-0-0-43, C 312, H. C. Peters Fonds, Archives of Ontario.

89. Allan to Edwards, October 8, 1911.

90. Allan to Edwards, October 27, 1911.

91. "Mine Timber for the Hollinger," *Porcupine Advance*, October 11, 1912.

92. S. Price, "Report re. Limitation of the Hours of Labor of Underground Workmen in the Mines of Ontario," 6.

93. During strikes in 1912, strikebreakers and strikers clashed repeatedly. "Strikers Fined for Assault," *Labor Gazette*, January 1913, 800. During Price's investigations, 332 (anonymous) ballots were returned in favor of the eight-hour day and only twelve against. "View of Employees," *Labor Gazette*, June 1913, 1392.

94. Price, "Report re. Limitation of the Hours," 6. On the 1899 miners strike in British Columbia over the eight-hour day, see Jeremy Mouat, "The Genesis of Western Exceptionalism: British Columbia's Hard Rock Miners, 1895–1903."

95. "View of Employees," *Labor Gazette,* June 1913, 1392.

96. Price, "Report re. Limitation of the Hours," 7.

97. Price, "Report re. Limitation of the Hours," 8–9.

98. "Commissioner Price's Report," *Canadian Mining Journal* 34, no. 9 (1913): 262.

99. *Hollinger Gold Mines Limited Annual Report,* January 1913, F113 Hollinger Company Fonds, Archives of Ontario, 25.

100. *Hollinger Gold Mines Limited First Annual Report,* January 1912, 25.

101. Harry Taylor to Edwards, May 23, 1913, MG 30 William S. Edwards Fonds, LAC.

102. *Hollinger Gold Mines Limited Annual Report,* January 1913, 26.

103. Thomas Gibson, "Introductory Letter," in *Annual Report of the Bureau of Mines, 1913,* 3, 1–2.

104. "Elk Lake Is Barring Out Indians," *Porcupine Advance,* August 16, 1912.

105. Cyril Young, Haileybury, to F. Pedley, Ottawa, "Application," April 28, 1911, RG 10, vol. 7460, file 18043-3 (Nippissing Agency).

106. J. D. McLean, Ottawa, to Cyril Young, Haileybury, May 8, 1911, RG 10, vol. 7460, file 18043-3 (Nippissing Agency).

107. Cyril Young to J. D. McLean, May 10, 1911, and J. D. McLean to Cyril Young, May 13, 1911, RG 10, vol. 7460, file 18043-3 (Nippissing Agency).

108. Thomas Gibson, "Statistical Review," in *Report of the Bureau of Mines, 1912* (Toronto: King's Printer, 1912), 9–10.

109. Thomas Gibson, "Statistical Review," in *Report of the Bureau of Mines, 1914* (Toronto: King's Printer, 1914), 9.

110. Gibson, "Statistical Review," in *Report of the Bureau of Mines, 1912,* 51.

111. E. T. Corkill, "Mines of Ontario," in *Report of the Bureau of Mines, 1912,* 154.

112. "Casualties," 298; "McIntyre and Jupiter," 668.

113. "Casualties," 298.

114. Company descriptions can be found in Bureau of Mines' annual reports for 1912 (152–58) and 1915 (130–44).

115. A. G. Burrows, "The Porcupine Gold Area," in *Twenty-Fourth Annual Report of the Bureau of Mines, 1915,* pt. 3 (Toronto: King's Printer, 1915), 3.

116. *Hollinger Gold Mines Ltd Annual Report 1914,* Archives of Ontario, 3.

117. "Report Recommending the Consolidation of Hollinger Gold Mines, Limited Acme Gold Mines, Limited Millerton Gold Mines, Limited Claim 13147 of Canadian Mining & Finance Co., Limited," *Hollinger Annual Report 1914,* Archives of Ontario, 2.

118. "Report Recommending the Consolidation of Hollinger Gold Mines," 2, 3.

119. "Report Recommending the Consolidation of Hollinger Gold Mines," 3.

120. "Report Recommending the Consolidation of Hollinger Gold Mines," 3.

121. "Report Recommending the Consolidation of Hollinger Gold Mines," 16–18.

122. "Two Young Men Lose Lives Fleeing Forest Fires Menacing Northern Towns," *Globe,* July 6, 1911.

123. Thomas Gibson, "Drama and Romance Form Background of Mining Progress," *Globe,* January 3, 1935; Gray, "Fire That Wiped Out Porcupine."

124. Karen Bachman, "Pioneers Share Fire Tales," *Timmins Daily Press,* July 8, 2011.

125. Diane Armstrong, "Remembering the Porcupine Fire," *Timmins Daily Press,* July 25, 2012.

Chapter 3. No Energy for Industry

1. Gold proved less desirable than other metals for the war effort and Porcupine benefited less from wartime demand than other mining zones. Further, as a mining frontier still under development, Porcupine remained dependent on inputs of labor and investment for profitability when the war began. As a general rule, "producing" mines with established infrastructure fared better than those still in exploration phases.

2. The basic structure of the mines—with ore mined from underground faces, conveyed to central shafts, and raised to the surface for processing in the mill—would largely remain the same until the end of the twentieth century. The scale of these operations would however increase after 1920. Petersen, "Origins of Canadian Gold Mining," 51, 436. See also Newell, *Technology on the Frontier.*

3. Both had been owned and managed by the Canadian Mining Finance Company before 1912.

4. A. G. Burrows, "The Porcupine Gold Area," in *Report of the Bureau of Mines, 1911* (Part II) (Toronto: King's Printer, 1911), 32.

5. Human alienation from nature under industrial economies has become a common trope in environmental history. However, some scholars suggest, as I do here, that the relationship between humans and nature is more complicated. Richard White argues that workers know land through labor. Richard White, *The Organic Machine: The Remaking of the Columbia River.* Liza Piper distinguishes between industrialization and commodification. She argues that commodification and the market economy, more so than industrialization, is what alienates people from the land. "Markets divorced industrial commodities from local places, even if industrialization remained closely tied to nature, and together shaped the long-term consequences of industrial operations to subarctic environments." Piper, *Industrial Transformation of Subarctic Canada*, 8.

6. *Twenty-Fourth Annual Report of the Ontario Bureau of Mines, 1915* (Toronto: King's Printer, 1915), 1.

7. *Annual Report of the Ontario Bureau of Mines, 1915*, 2.

8. "The Cost of Production," *Northern Miner*, October 30, 1915, 1.

9. Gold production jumped from $42,637 in 1911 to $2,114,086 in 1912, then doubled again in 1913 to $4,558,518. So the 21 percent gain to $5,529,767 might have appeared modest in comparison. *Annual Report of the Ontario Bureau of Mines, 1915*, 5.

10. *Annual Report of the Ontario Bureau of Mines, 1915*, 8.

11. "Gold Mining," *Canadian Mining Journal* (September 1, 1914): 567.

12. *Annual Report of the Ontario Bureau of Mines, 1915*, 8.

13. "Porcupine, Kirkland Lake, and Sesikinika," *Canadian Mining Journal* (August 15, 1914): 563.

14. Foley O'Brien suffered the additional misfortune of a fire in July of 1914. Ben Hughes, "Gold Mining at Porcupine and Kirkland Lake, Ontario," *Canadian Mining Journal* (October 1, 1914): 635.

15. A. G. Burrows, "The Porcupine Gold Area," in *Twenty-Fourth Annual Report of the Ontario Bureau of Mines, 1915*, pt. 3 (Toronto: King's Printer, 1915), 20–41.

16. "Special Correspondence Porcupine, Kirkland Lake and Sesikinika," *Canadian Mining Journal* (September 1, 1914): 592.

17. "Special Correspondence Porcupine, Swastika, Kirkland Lake," *Canadian Mining Journal* (October 15, 1914): 689.

18. *Fifth Annual Report Covering Operations for the Year 1915, Hollinger Gold Mines, Limited* (Toronto, 1916), 2.

19. *Twenty-Fifth Annual Report of the Ontario Bureau of Mines, 1916* (Toronto: King's Printer, 1916), 1; "The Working of Small Ore Deposits," *Canadian Mining Journal* (April 1, 1915): 197.

20. *Fifth Annual Report Governing Operations for the Year 1915 Hollinger Gold Mines Limited*, 23.

21. "Gold Mining," *Canadian Mining Journal* (September 1, 1914): 567. An unidentified alternate source was found by November 1914. "Cobalt, Elk Lake, Gowganda, South Lorraine," *Canadian Mining Journal* (November 1, 1914): 722. Russell Ally discusses the relationship between Britain and South Africa in *Gold & Empire: The Bank of England and South Africa's Gold Producers, 1886–1926*, 29–46.

22. *Twenty-Fifth Annual Report of the Ontario Bureau of Mines, 1916* (Toronto: King's Printer, 1916), 7.

23. *Sixth Annual Report Governing Operations for the Year 1916, Hollinger Consolidated Gold Mines* (Toronto, 1917), 2, 17.

24. Petersen, "Origins of Canadian Mining," 24.

25. *Sixth Annual Report Governing Operations for the Year 1916, Hollinger Consolidated Gold Mines*, 2, 17.

26. *Annual Report Governing Operations for the Year 1917, Hollinger Consolidated Gold Mines* (Toronto, 1918), 12.

27. *Sixth Annual Report Governing Operations for the Year 1916, Hollinger Consolidated Gold Mines*, 17.

28. Petersen, "Origins of Canadian Gold Mining," 297.

29. Petersen, "The Origins of Canadian Gold Mining," 297.

30. *Twenty-Ninth Annual Report of the Ontario Department of Mines, 1920* (Toronto: King's Printer, 1920), 86.

31. *Sixth Annual Report Governing Operations for the Year 1916, Hollinger Consolidated Gold Mines*, 17.

32. "Men Back Again at Work at McIntyre," *Porcupine Advance*, March 6, 1918.

33. With some exceptions. The lack of labor influenced the adoption of Glory Holes at Dome, Millerton, and Hollinger by 1917. Glory Holes involved running a tunnel under the gold deposit, blasting holes in the ceiling, and slowly funneling the ore into waiting cars in the tunnels using gravity. The technique could cause major collapses and large holes on the surface. "Glory Hole at Hollinger," *Northern Miner*, April 14, 1917, 3.

34. *Sixth Annual Report Governing Operations for the Year 1916, Hollinger Consolidated Gold Mines*, 9.

35. "1917 Summary—Mines Did Well," *Northern Miner*, January 5, 1918, 2; *Twenty-Seventh Annual Report of the Ontario Bureau of Mines, 1918*, vol. 27, pt. 1 (Toronto: King's Printer, 1918), 5.

36. "1917 Summary—Idle Properties," *Northern Miner*, January 5, 1918, 2.

37. *Annual Report of the Ontario Bureau of Mines, 1918*, 6.

38. "1917 Summary—Dome" and "1917 Summary—Hollinger," *Northern Miner,* January 5, 1917, 5.

39. On the meaning and significance of ethnic labor at Porcupine, see Petersen, "Origins of Canadian Gold Mining," 422–33.

40. Abel, *Changing Places,* 108.

41. For example, see Peter Vasiliadis, *Dangerous Truth: Interethnic Competition in a Northeastern Ontario Goldmining Center.*

42. In *Capital,* volume 1, Karl Marx described this "nomad population," as "the light infantry of capital, thrown by it, according to its needs, now to this point, now to that."

43. Barnes, *Fortunes in the Ground,* 66.

44. Canada, Statistics Canada, 1911 Census, Ottawa, 1911; Canada, Statistics Canada, 1921 Census, Ottawa, 1921.

45. In 1914, twenty-one of thirty-eight; 1915, eleven of twenty-one; 1916, twenty-nine of fifty-one; 1917, sixteen of thirty-six; 1818, twelve of fourteen; and 1919, twenty-one of thirty-nine mine accident fatalities were non-English-speaking people.

46. E. T. Corkill, "Mining Accidents," in *Annual Report of the Bureau of Mines, 1911,* 66.

47. Eleven out of twenty-one men killed in or around the mines in 1915 were non-English speaking. T. F. Sutherland, "Mining Accidents in Ontario in 1915," in *Annual Report of the Ontario Bureau of Mines, 1916,* 55; "All Alien Enemies Discharged," *Northern Miner,* November 13, 1915, 1.

48. Resulting in the British Non-ferrous Metal Industry Act in 1918. Report of the Departmental Committee Appointed by the Honourable Martin Burrell, Federal Minister of Mines, to Consider the British Non-Ferrous Metal Industry Act, 1918, and to Advise with Respect to Similar Legislation for Canada (Ottawa, 1918), No. 77: List of Minerals in Short Supply, RG 87 Mineral Resources Branch, LAC, viii.

49. Report of the Departmental Committee, xi–xiii.

50. Royal commission on the natural resources, trade, and legislation of certain portions of His Majesty's dominions; minutes of evidence taken in central and western provinces of Canada in 1916, pt. 2, appendix 7, "Porcupine Gold Mining District of Ontario" (London: His Majesty's Stationery Office, 1917), 307.

51. "Italians for the Gold Mines," *Northern Miner,* October 9, 1920, 2; Howard Poillon to Henry Depencier, October 30, 1920.

52. A. H. Paterson, "Mining at the McIntyre Porcupine Mines," Summer Essay 1924, file 8, Student Essays, School of Mining and Agriculture Fonds, Queen's University Archives, Kingston, Ontario, 22.

53. "Government Aid Urged for Gold Mines," *Porcupine Advance,* July 17, 1918. The movement eventually failed. "Dominion Government Not to Aid Gold Mines," *Porcupine Advance,* August 14, 1918; "Western Gold Production Falling," *Canadian Mining Journal* (August 1, 1918): 267; "Spokane—Aug. 26," *Engineering and Mining Journal,* September 7, 1918, 467.

54. Petersen, "Origins of Canadian Gold Mining," 309.

55. Newell, *Technology on the Frontier,* 16–17.

56. "Special Correspondence," *Canadian Mining Journal* (March 15, 1912): 209.

57. "Porcupine Honeycombed by Diamond Drilling," *Porcupine Advance,* July 19, 1916.

58. "Will Systematically Prospect Dome Ex. For Dome Ore Body," *Porcupine Advance,* June 21, 1916.

59. W. B. Baker, "The Geology of Kingston and Vicinity," in *Twenty-Fifth Annual Report of the Bureau of Mines, 1916*, vol. 25, pt. 3 (Toronto: King's Printer, 1916), 1.

60. "Personal and General," *Canadian Mining Journal* (July 15, 1914): 493.

61. A. G. Burrows, "The Porcupine Gold Area," in *Annual Report of the Bureau of Mines, 1915*, pt. 3, 29.

62. Burrows, "The Porcupine Gold Area," in *Annual Report of the Bureau of Mines, 1915*, pt. 3, 30.

63. *Annual Report of the Ontario Bureau of Mines, 1916*, 251, 261.

64. A. G. Burrows, "The Porcupine Gold Area," in *Annual Report of the Ontario Bureau of Mines, 1915*, pt. 3, 44, 47.

65. *Twenty-Seventh Annual Report of the Ontario Bureau of Mines, 1918* (Toronto: King's Printer, 1918), 5.

66. "Unexplored Canada," *Northern Miner*, February 3, 1917, 2.

67. "Prospecting—War Affects Prospectors," *Porcupine Advance*, March 7, 1917.

68. "The War Has Taken a Big Proportion, and the Mines Have Taken Many" and "Prospectors Not So Plentiful Now," *Porcupine Advance*, March 7, 1917.

69. "Shining Tree—Area Suffering A Partial Eclipse, Discontinuance of Work on Some Properties," *Porcupine Advance*, March 21, 1917.

70. "Posts for Prospectors: Stampeders vs. Prospectors," *Northern Miner*, September 18, 1915, 8.

71. "The Rush into Kow Kash," *Northern Miner*, September 11, 1915, 4.

72. He suggests that a low, timbered area might be converted into a plant: "Of course, the timber has been burned some and therefore dead, but it is good wood. No trouble to blow out a hole for water." A neighboring prospector had a road in place which might be used to access the area. H. C. Anchor to W. S. Edwards, "Dome Ex," November 16, 1915, 2-27 Correspondence, general business, 1915, MG 30 W. S. Edwards fonds, LAC.

73. H. C. Anchor to W. S. Edwards, November 29, 1915, 2-27 Correspondence, general business, 1915, MG 30 W. S. Edwards fonds, LAC.

74. H. C. Anchor to W. S. Edwards, "Re. Lot 7," June 6, 1916, 2-27 Correspondence, general business, 1915, MG 30 W. S. Edwards fonds, LAC.

75. H. C. Archer to W. S. Edwards, December 9, 1918, 2-27 Correspondence, general business, 1915, MG 30 W. S. Edwards fonds, LAC.

76. "Prospectors vs. Engineers," *Northern Miner*, November 4, 1916, 3.

77. "Prospecting," *Northern Miner*, January 27, 1917, 6.

78. "Inexperience Can Condemn a Camp," *Porcupine Advance*, June 4, 1915.

79. "Prospectors vs. Engineers," *Northern Miner*, November 4, 1916, 3.

80. "Mining Future of Ontario," *The Northern Miner*, April 28, 1917, 1, 5.

81. "Developing a Career: Old Hand Gives Advice to the Young Mining Engineer," *Northern Miner*, May 22, 1920.

82. Minute Book, Board of Governors of the School of Mining and Agriculture Dec. 1892–Nov. 1909, file 3, box 1, Queen's University School of Mining 1893–1914 Ledgers, School of Mining and Agriculture Fonds, Queen's University Archives, Kingston, Ontario.

83. Preserved correspondence ends in 1911, but the letters suggest a standing working relationship between the school and the mines that probably extended further

back. For example, see E. H. Calling to E. C. Keeley, March 13, 1911, file 6, Correspondence to Dr. Goodwin, 1911, Jan–June A–H, School of Mining and Agriculture Fonds, Queen's University Archives, Kingston, Ontario: "There are so many applicants for the job at Porcupine that I am leaving the election of the men with Prof. Guillan and Dr. Goodwin."

84. B. H. Budgeon, "Mining Methods at the Hollinger Consolidated Gold Mines," Summer Essay 1923, file 8, Student Essays, School of Mining and Agriculture Fonds, Queen's University Archives, Kingston, Ontario.

85. "After the war the explorers will scatter all through these wildernesses." "Unexplored Canada," *Northern Miner,* February 3, 1917, 2.

86. "Prospecting a Dying Industry; Govt Indifference a Big Cause," *Northern Miner,* January 1, 1921, 3.

87. "Classes May Be Held Here For Prospectors," *Porcupine Advance,* November 24, 1920.

88. "Classes for Prospectors Make Good Start Here," *Porcupine Advance,* May 4, 1921.

89. Lorena Campuzano Duque makes a similar argument about artisanal miners and industrial regimes in Antioquia, where industrialization depended on the knowledge and labor of vernacular miners. See chapters 3–4 in "The People's Gold: Race and Vernacular Mining in the 'Ailing' Landscapes of Antioquia, Colombia, 1540–1958."

90. See Ian Mosby, *Food Will Win the War: The Politics, Culture, and Science of Food on Canada's Home Front.*

91. In *Changing Places,* Kerry Abel writes of the area around Timmins that "working life on the farm was clearly a struggle." She describes the short growing season, poor markets, lack of transportation access, lack of labor, and environmental disaster (especially the 1916 fires) that saw "lands abandoned almost as quickly as they were taken up." Abel, *Changing Places,* 154. J. David Wood also addresses the failure of agriculture settlement in northern Canada (including the Porcupine/Timmins region) in *Places of Last Resort.* He writes, "The opening of new land for farming after 1910 and the rush of people into that land present a saga of human desire and elusive opportunity. By most measures, the occupation of the marginal areas during the inter-war period was widely blighted by failure." Wood, *Places of Last Resort,* 179.

92. A. G. Burrows, "The Porcupine Gold Area," in *Annual Report of the Bureau of Mines, 1915,* pt. 3, 1–2.

93. Henry Depencier to Major W. J. Morrison, November 24, 1920, file: "Introductions," box 1, F 1250, Dome Mining Company Fonds, Archives of Ontario.

94. "Scheme to Develop the North Country," *Porcupine Advance,* April 12, 1916.

95. "Free Lots Offered Here for Greater Production," *Porcupine Advance,* May 23, 1917; "Cultivate a Lot and Help a Little," *Porcupine Advance,* May 30, 1917; "New Experimental Farm for the Great North Land," *Porcupine Advance,* February 28, 1917.

96. Potatoes had been successfully grown by Indigenous people since at least the turn of the century, as seen in the early Ontario Bureau of Mines reports. "Honey Bees Thrive in the North Land," *Porcupine Advance,* February 28, 1917; "Results of Potato Culture Experiments by the Government," *Porcupine Advance,* May 17, 1916.

97. "Flax as Profitable Crop for North Land," *Porcupine Advance,* March 7, 1917.

98. According to the Honorable Manning Doherty, "The great agricultural future of the North Country lay in livestock rather than grain." "To Test Northland as a Ranching Country," *Porcupine Advance,* August 11, 1920.

99. "Every Little Will Help in Production of Food," *Porcupine Advance,* May 2, 1917.

100. "Strawberries and Flowers in Winter in the North Land," *Porcupine Advance,* May 2, 1917; "One of Mountjoy's Successful Farmers," *Porcupine Advance,* June 26, 1918; "Mrs Parsons Suggests Food Allowance," *Porcupine Advance,* September 12, 1917.

101. "Importance of Mining," *Northern Miner,* December 16, 1916, 2.

102. "Re-opening Old Properties," *Northern Miner,* January 3, 1920, 4; "They Are Coming Back," *Northern Miner,* October 9, 1920, 2.

103. "The Gold Camps of Northern Ontario," *Northern Miner,* November 23, 1918, 1.

104. "Dome Lake Mines Report, 1919," *Northern Miner,* March 27, 1920.

105. "Teck-Hughes Gold Mine—Workers Quit Temporarily over Food," *Porcupine Advance,* July 10, 1918; "Strikes and Lockouts," *Porcupine Advance,* July 23, 1919; "Porcupine Miners' Union—Vote on Striking," *Porcupine Advance,* May 23, 1923.

106. "Settlement in Porcupine," *Northern Miner,* July 26, 1919, 1.

107. Howard Poillon to Henry Depencier, October 30, 1920, file: "Introductions," box 1, Dome Mines Company Fonds, F 1350, Archives of Ontario.

108. "Report of Hollinger Officers," *Northern Miner,* February 6, 1920, 8.

109. "Ontario has Only Flourishing Gold Field," *Northern Miner,* August 7, 1920, 4.

110. "World Needs More Precious Metals," *Northern Miner,* February 28, 1920, 3.

111. Geo. Wright to Henry Depencier, January 3, 1923, Dome Fonds, box 1, Exploration, 1923, Archives of Ontario; Noah Timmins, "Director's Report," *Annual Report for Hollinger Consolidated Gold Mines, Limited,* February 1, 1923, Hollinger Company Fonds, Hollinger Annual Reports, 1912–35, F 1335, Archives of Ontario, 3.

112. Jean L. Manore, *Cross-Currents: Hydroelectricity and the Engineering of Northern Ontario,* 43–44.

113. "Porcupine, Kirkland Lake, and Harricanaw," *Canadian Mining Journal* (January 1, 1914): 30; "Water Powers and Mines," *Northern Miner,* December 4, 1915, 2; *Annual Report of the Ontario Bureau of Mines, 1916,* 5; "More Power for Porcupine," *Northern Miner,* August 26, 1916, 7.

114. "Power Available at Porcupine, Storage Most Vital Point," *Northern Miner,* February 26, 1916, 7.

115. "Power Troubles in Porcupine," *Northern Miner,* April 14, 1917, 3.

116. "Miller Lake—Break in Dyke Floods School Buildings," *Porcupine Advance,* March 21, 1917.

117. "Miller Lake—Dam Breaks Again," *Porcupine Advance,* April 4, 1917.

118. The estimate was based on the capacity of the plants, not actual usage, which was only 8,000 horsepower even two years later.

119. Statement of Appellants' Case, on Appeal from the Supreme Court of Ontario Appellate Division, file 57-13, *Northern Canada Power Co. et al. v. Hollinger Consolidated Gold Mines et al.,* P.C. 1924, MG 28 III 35 57, LAC.

120. "Power Supply at Porcupine Must Be Increased," *Northern Miner,* December 25, 1920, 1.

121. J. H. Black to Department of Indian Affairs, June 4, 1921, RG 10 (C11559), vol. 7584, file 6065-3, Dams and Flooding on Mattagami, 1921–38, LAC.

122. Memo, H. J. Bury to Acting Deputy Superintendent General, "Raise of Level of Mattagami River," August 5, 1921, RG 10 (C11559), vol. 7584, file 6065-3, Dams and Flooding on Mattagami, 1921–38, LAC.

123. Jas. Naveau (Chief) to the Minister, Department of Indian Affairs, October 1, 1921, RG 10 (C11559), vol. 7584, file 6065-3, Dams and Flooding on Mattagami, 1921–38, LAC; Deputy Superintendent General to Chief James Navion, October 6, 1921, RG 10 (C11559), vol. 7584, file 6065-3, Dams and Flooding on Mattagami, 1921–38, LAC.

124. Luby, *Dammed;* Manore, *Cross-Currents,* 53.

125. PC 2523, December 30, 1919, Correspondence Regarding Indians and Mining in General, 1919–37, RG 10 (C-11587), file 18001, pt. 1, vol. 7630, LAC.

126. See, for example, J. Lorn McDougal to Duncan Campbell Scott, November 17, 1922, "Abitibi Mining," vol. 7632, file 18074, RG 10 (C11589), LAC. The letter is a request to explore and stake nine hundred acres on the Abitibi Indian Reserve No. 70, just east of the Porcupine mining area. Several similar letters exist in this file.

127. Department of Indian Affairs rebuffed applications and arguments for claim-staking on reserve before the First World War. See discussion of Cyril T. Young in chapter 2.

128. A. H. Black to Department of Indian Affairs, December 30, 1921, RG 10 (C11559), vol. 7584, file 6065-3, Dams and Flooding on Mattagami, 1921–38, LAC.

129. Asst. Deputy and Secretary J. D. McLean to Northern Canada Power, February 13, 1922, RG 10 (C11559), vol. 7584, file 6065-3, Dams and Flooding on Mattagami, 1921–38, LAC.

130. His house had been big and well appointed (including an organ, which was destroyed) and attached to an acre of land on which he grew potatoes. Godfrey to J. D. McLean, May 16, 1923, RG 10 (C11559), vol. 7584, file 6065-3, Dams and Flooding on Mattagami, 1921–38, LAC.

131. T. J. Godfrey to Department of Indian Affairs, September 22, 1923, RG 10 (C11559), vol. 7584, file 6065-3, Dams and Flooding on Mattagami, 1921–38, LAC.

132. Northern Canada Power to Department of Indian Affairs, October 18, 1923, RG 10 (C11559), vol. 7584, file 6065-3, Dams and Flooding on Mattagami, 1921–38, LAC.

133. Indian Agent, Sturgeon Falls, to Department of Indian Affairs, June 27, 1925, RG 10 (C11559), vol. 7584, file 6065-3, Dams and Flooding on Mattagami, 1921–38, LAC.

134. Naveau and his council wrote the Department of Indian Affairs in 1923 to ask what would be done but received no response. James Naveau, Sam Luke, and James Naveau to Department of Indian Affairs, August 17, 1923, RG 10 (C11559), vol. 7584, file 6065-3, Dams and Flooding on Mattagami, 1921–38, LAC. They wrote again in 1925. Mattagami Band to Department of Indian Affairs, September 30, 1925, RG 10 (C11559), vol. 7584, file 6065-3, Dams and Flooding on Mattagami, 1921–38, LAC.

135. Fasken, Robertson, Sedgewick, Aitchison & Pickup to Messrs. Blake & Redden, "Re: Northern Canada v. Hollinger," May 18, 1925, file 57-13, Northern Canada Power Co. et al. v. Hollinger Consolidated Gold Mines et al. P.C. 1924, Correspondence 1924–26, MG 28 III 3557, LAC.

136. Hollinger agreed to draw power from a new Northern Canada Power plant on the Quinze River in Quebec. Northern Canada Power eventually paid $500,000

in compensation. Hollinger absorbed the cost of a now-useless transmission line. "Is Price Cut in Mine Power Coming?," *Northern Miner,* April 3, 1926; Manore, *Cross-Currents,* 50–51.

137. "Government's Inaction Criticized by Mac Lang," *Porcupine Advance,* February 28, 1923.

138. "The Porcupine—The Golden-Hearted Land," *Porcupine Advance,* December 20, 1922.

139. "Movies of the Hollinger," *Northern Miner,* March 6, 1920, 5.

Chapter 4. Mine Waste

1. "Real Canada—Uninviting Place," *Darling Downs Gazette,* May 12, 1911, retrieved from the National Archives of Australia's Trove, http://nla.gov.au/nla.news-article182678881.

2. LeCain, *Mass Destruction,* 11–12, 128.

3. Andy Horowitz, *Katrina: A History, 1915–2015,* 12–16.

4. Piper, *Industrial Transformation of Subarctic Canada,* 143–44.

5. Price, "Report re. Limitation of the Hours," 11–12.

6. Robert N. Chester III makes a similar argument about Comstock in "Consequences of the Comstock: The Remaking of Working Environments on America's Largest Silver Strike, 1859–1880," in *Mining North America,* ed. McNeil and Vrtis, 109.

7. Kevin Vincent self-published a volume called *Bootlegged Gold: Amazing Untold Stories from the 20th Century's Lucrative World of Gold Smuggling* that gives a somewhat sensationalized summary of the high-grading problem in Porcupine. See also "A. Adjison Guilty of Highgrading," *Porcupine Advance,* January 18, 1922; "Three Year Term for One High-Grader Last Week," *Porcupine Advance,* May 27, 1925; "Charged with Illegal Possession of Gold Ore," *Porcupine Advance,* June 10, 1926; "Not Less Than a Year on Charge of High-Grading," *Porcupine Advance,* July 7, 1927; "Fine of $1500 Imposed, No High Grading Charge," *Porcupine Advance,* October 13, 1927; "Jail Terms Given in Two Gold Ore Cases," *Porcupine Advance,* December 1, 1927.

8. Thomas Johns to Henry Depencier, November 5, 1921, box 3 MU 8689, K. S. and A. F. Reports, 1922, Dome Mines Company Fonds, F 1350, Archives of Ontario.

9. Petersen, "Origins of Canadian Gold Mining," 437.

10. A. F., South Porcupine, January 15, 1922.

11. Slime referred to crushed tailings getting mixed with chemicals to extract fine gold. A. F., South Porcupine, February 3, 1922.

12. A. F., South Porcupine, February 4, 1922.

13. A. F., South Porcupine, February 12, 1922.

14. For a discussion of risk, labor, and mining in Canada in the early twentieth century, see Karen Buckley, *Danger, Death, and Disaster in the Crowsnest Pass Mines, 1902–1928.*

15. A. F., South Porcupine, March 18, 1922.

16. A. F., South Porcupine, May 26 and 27, 1922.

17. A. F. wrote, "At 2.00 A.M. a workman was crushed by falling rock on #10 level." A. F., South Porcupine, May 27, 1922.

18. A mining tradition. See, for example, Gail Weir, *The Miners of Wabana: The Story of the Iron Ore Miners of Bell Island,* 146–47.

19. A. F., South Porcupine, May 27, 1922.

20. The industry and the government tended to worry about miner deaths in a broader statistical sense. Individual deaths were not in themselves particularly interesting, but trends (increases or decreases) warranted official attention. See, for example, T. F. Sutherland, "Report on the Mining Accidents in Ontario in 1925," *Bulletin No. 54* (Toronto: King's Printer, 1926), 1. Of course, dissatisfied workers could also interrupt production, and so the decision to keep working after the deaths had to be carefully weighed against the possibility of a strike. Dome followed the example of its international colleagues in the decision to continue work. In the aftermath of three deadly explosions in Colorado Coal Mines in 1910, the mines reopened as soon as logistically possible, despite the discomfort of workers and community members. Thomas Andrews describes this decision sparked discomfort among miners and community members and "reenergized long-festering conflicts among mine workers, mining companies, and the broader public." Thomas Andrews, "Dust to Dust: The Colorado Coal Mine Explosion Crisis of 1910," in *Mining North America*, ed. McNeill and Vrtis, 133.

21. Dowell saw his interests as aligning more closely with the wider industrial workforce to the south than with managers working in the same industry. His identity as "miner" was less powerful than his identity as "worker." For the latter half of the nineteenth century, Arn Keeling and Patricia Boulter note that miners tended to form powerful identities around their occupation. This does not necessarily contradict the evidence from the A. F. reports, but may represent the fact that, in the 1920s, the identity of the industrial miner was still in the making. Arn Keeling and Patricia Boulter, "From Igloo to Mine Shaft: Inuit Labor and Memory at the Rankin Lake Nickel Mine."

22. A. F., South Porcupine, May 29, 1922.

23. A. F., South Porcupine, May 15, 1922.

24. Royal commission on the natural resources, trade, and legislation of certain portions of His Majesty's dominions; minutes of evidence taken in central and western provinces of Canada in 1916, pt. 2, appendix 7, Porcupine Gold Mining District of Ontario (London: His Majesty's Stationery Office, 1917), 307.

25. A. F., South Porcupine, February 23, 1922.

26. A. F., South Porcupine, January 29, 1922.

27. "Hollinger Tailings Pond," *Northern Miner*, April 8, 1916, 4.

28. "New Dome Road Almost Finished," *Porcupine Advance*, February 16, 1916.

29. "Tails in Dispute," *Northern Miner*, December 15, 1917, 1.

30. Examination—H. P. Depencier in the Supreme Court of Ontario between the Digby Dome Mines Company Limited and the Dome Mines Company Limited, January 15, 1923, 3, Digby Vet (Ontario Supreme Court Case), General Superintendent's Files, 1926–35, box 6, Dome Mines Company Fonds, F 1350, Archives of Ontario.

31. Neither the lawyers nor Depencier really understood this process, as exhibited by their back-and-forth during this part of the case. Frustrated, Depencier eventually exclaimed that "it is just a little higher and a little higher and then commences to silt across on the other side. Possibly the wind has an effect on it, I don't know." Depencier's interest in ore ended at the mill. Examination—H. P. Depencier, January 15, 1923, 7–8.

32. Examination—H. P. Depencier, January 15, 1923, 12, 13, 15.

33. Letter to S. B. Celement, Esq., Chief Engineer, T&NO Railway, September 23, 1922, Tailings, box 3, Dome Mines Company Fonds, F 1350, Archives of Ontario.

34. Statement of Facts on Behalf of Dome in the Supreme Court of Ontario between Digby Dome Mines Company Limited and Dome Mines Company Limited, 4, Tailings, Digby, box 10, General Superintendent's Files, 1926–35, Dome Mines Company Fonds, F 1350, Archives of Ontario.

35. According to Dome's statement of facts, the price may have been based on a similar transaction recently conducted by Hollinger where the company had in fact bought "one or two small parcels that were necessary to the control of the area" for an average price of about $2000. Statement of Facts on Behalf of Dome, 5.

36. Examination—H. P. Depencier, January 15, 1923, 17–18, 20, 22.

37. Alex Fasken to Depencier, October 26, 1922, Tailings, box 3, Dome Mines Company Fonds, F 1350, Archives of Ontario.

38. "The crown insists on insertion in the grant to your Company of a reservation. . . . The result of this will be that you will never be able to drain the lake or to foul the waters of the lake without settling with all riparian owners." Alex Fasken to Henry Depencier, "Re. Purchase of the Bed of Porcupine Lake," September 9, 1924, Tailings, 1923–24, box 10, Dome Mines Company Fonds, F 1350, Archives of Ontario.

39. Statement of Facts on Behalf of Digby, in the Supreme Court of Ontario Between Digby Dome Mines Company Limited and Dome Mines Company Limited, 19, Tailings, Digby Vet, box 10, Dome Mines Company Fonds, F 1350, Archives of Ontario.

40. Lawrence and Davies, "Sludge Question"; Carolyn Merchant, *Green Versus Gold: Sources in California's Environmental History;* Jeremy Manuel, "Efficiency, Economics, and Environmentalism: Low Grade Iron Ore Mining in the Lake Superior District, 1913–2010," in *Mining North America,* ed. McNeill and Vrtis, 191–216; Robynne Mellor, "A Comparative Case Study of Uranium Mine and Mill Tailings Regulation in Canada and the United States," in *Mining North America,* ed. McNeill and Vrtis, 256–79; Fredric Quivik, "Smoke and Tailings: An Environmental History of Copper Smelting Technologies in Montana, 1880–1930"; Marilyn Ziebarth, "California's First Environmental Battle."

41. Caitlynn Beckett and Arn Keeling, "Rethinking Remediation: Mine Reclamation, Environmental Justice, and Relations of Care."

42. Report of the Hollinger Mine Inquiry, Government Commissions file, Record Group 18, B-82, 1–5, Archives of Ontario.

43. "Hollinger System Blamed," *Northern Miner,* May 31, 1928.

44. "Hollinger Probe Opens," *Northern Miner,* March 1, 1928.

45. "Hollinger System Blamed."

46. "Hollinger System Blamed."

47. "Perished in Fire," *Mail,* February 11, 1928, retrieved from National Library of Australia's Trove, http://nla.gov.au/nla.news-article58549268; "Trapped: Fire in Mine, Too Many for Lifts," *Sun* (Sydney), February 11, 1928, retrieved from National Library of Australia's Trove, http://nla.gov.au/nla.news-article224223232.

48. "Hollinger System Blamed."

49. Hearings Before the Committee on Mines and Mining United States Senate Seventeenth Congress Second Session on S. 2079: *A Bill Authorizing an Appropriation for Mining Experiment Stations of the United States Bureau of Mines,* pt. 2, February 15, 1929 (Washington, DC: Government Printing Office, 1929), 30–31.

50. "Hollinger System Blamed."

51. "Disaster Befalls Men in Hollinger Mine," *Daily Colonist,* February 11, 1928, retrieved from the University of Victoria's Daily Colonist Digital Collection, https://archive.org/details/dailycolonist0228uvic_9; "40 Canadian Miners Are Trapped," *Madera Tribune,* February 11, 1928, retrieved from California Digital Newspaper Collection, https://cdnc.ucr.edu/cgi-bin/cdnc?a=d&d=MT19280211.2.2&srpos=4&e=-------en--20--1--txt-txIN-Hollinger+Fire-------1.

52. "Perished in Fire," *Mail,* February 11, 1928, retrieved from National Library of Australia's Trove, http://nla.gov.au/nla.news-article58549268; "Trapped: Fire in Mine."

53. The *Aukland Star* received a cable from Vancouver that knew the numbers of trapped men and the details of the rescue effort. Although it was rough on location, saying that the disaster had happened at North Bay. The *Otago Daily Times* had a cable from Sudbury with similar details, and so did the *Stratford Evening Post,* the *New Zealand Herald,* and other papers. "Forty in Peril," *Aukland Star,* February 13, 1928, retrieved from National Library of New Zealand's Paper's Past, http://paperspast.natlib.govt.nz/newspapers/AS19280213.2.46; "Fire in Canadian Mine," *Otago Daily Times,* February 13, 1928, retrieved from National Library of New Zealand's Paper's Past, http://paperspast.natlib.govt.nz/newspapers/ODT19280213.2.50; "Trapped in Mine: Canadian Tragedy," *Stratford Evening Post,* February 13, 1928, retrieved from National Library of New Zealand's Paper's Past, http://paperspast.natlib.govt.nz/newspapers/STEP19280213.2.30; "Fire in Goldmine," *New Zealand Herald,* February 13, 1928, retrieved from National Library of New Zealand's Paper's Past, http://paperspast.natlib.govt.nz/newspapers/NZH19280213.2.65.

54. "The Quartz Mine Fire," *Otago Daily Times,* February 15, 1928, retrieved from National Library of New Zealand's Paper's Past, http://paperspast.natlib.govt.nz/newspapers/ODT19280215.2.90.

55. "Hollinger Mine," *Otago Daily Times,* March 26, 1928, retrieved from National Library of New Zealand's Paper's Past, http://paperspast.natlib.govt.nz/newspapers/ODT19280326.2.38; "Gross Negligence," *Aukland Star,* March 26, 1928, retrieved from National Library of New Zealand's Paper's Past, http://paperspast.natlib.govt.nz/newspapers/AS19280326.2.83; "Mining Fire Disaster," *Northern Star,* March 26, 1928, retrieved from National Library of Australia's Trove, http://nla.gov.au/nla.news-article93667669.

56. "Mass Meeting Addressed by Speakers in Six Languages," *Porcupine Advance,* February 16, 1928.

57. Abel, *Changing Places,* 142.

58. T. F. Sutherland, *Report on the Mining Accidents in Ontario in 1928* (Toronto: King's Printer, 1929), 15.

59. "Latest Equipment for Combating Fire Installed at Mine," *Globe,* August 4, 1928, 2.

60. A strategy repeated elsewhere. See, for example, Jörg Arnold, "'The Death of Sympathy:' Coal Mining, Workplace Hazards, and the Politics of Risk in Britain, ca. 1970–1990." Arnold argues that miners' political bargaining power was eroded when their "conservative opponents" turned the miners' own vulnerabilities against them, arguing that "the very hazardous working conditions were taken as proof of an obstinate refusal of the industry to go with the times. The real danger, they argued, were not health hazards but the miners themselves."

61. Nancy Forestell, "And I Feel Like I'm Dying from Mining for Gold: Disability, Gender, and the Mining Community, 1920–1950"; Thomas Klubock, "Working-Class Masculinity, Middle-Class Morality, and Labor Politics in the Chilean Copper Mines," 435–37.

62. Forestell, "And I Feel Like I'm Dying from Mining for Gold," 89.

63. "Hollinger System Blamed."

64. "Hollinger System Blamed."

65. "Gross Negligence."

66. For example, "Trapped: Fire in Mine."

67. Forestell, "And I Feel Like I'm Dying," 90.

68. "Mining Disaster Fire in a Coal Mine," *Barrier Miner,* February 13, 1928, retrieved from National Library of Australia's Trove, http://nla.gov.au/nla.news-article46003426; "Hollinger Mine Fire," *Advocate,* retrieved from National Library of Australia's Trove, http://nla.gov.au/nla.news-article67572798. In fact, underground fires at gold mines were not totally unknown. Periodic reports appear in Australia and California. "Gold Mine Fire Out after Three Months," *Los Angeles Herald,* December 22, 1919, retrieved from California Digital Newspaper Collection, https://cdnc.ucr.edu/cgi-bin/cdnc?a=d&d=LAH19191222.2.1046&srpos=2&e=-------en--20--1--txt-txIN-underground+Gold+Mine+Fire-------1; "Gold Mine on Fire," *Evening News* (Sydney), March 20, 1903, retrieved from National Library of Australia's Trove, http://nla.gov.au/nla.news-article113403447.

69. Daniel Harrington, "Progress in Mine Ventilation," 9, 1, 10–12.

70. Daniel Harrington, "Fires in Metal Mines: Causes, Prevention, and Methods of Handling," *Canadian Mining Journal* (December 21, 1928): 1059–60.

71. Daniel Harrington, "Work of the Safety Division of the United States Bureau of Mines, Fiscal Year 1930."

72. Daniel Harrington, "Progress in Metal-Mine Ventilation in 1930," 5.

73. "Educating the People to Forest Conservation," *Porcupine Advance,* August 12, 1926.

74. S. A. Pain, *Three Miles of Gold: The Story of Kirkland Lake,* 69.

75. "Bayside Beach Resort Opened at Barber's Bay," *Porcupine Advance,* June 13, 1929.

76. "Special Features of the Big Dog Race, March 6th," *Porcupine Advance,* February 21, 1929; "Clair Severt Starts to Ski to Ottawa in 14 Days," *Porcupine Advance,* February 21, 1929; "Kirkland Lake Wins from McIntyre at Timmins 15–8," *Porcupine Advance,* July 24, 1930; "Successful T.B.A.A. Sports Day Here on Dominion Day," *Porcupine Advance,* July 3, 1930.

77. "Will You Help Save a Place in the North for Waterfowl?," *Porcupine Advance,* April 22, 1925.

78. "Efforts to Preserve the Migratory Birds," *Porcupine Advance,* March 29, 1929.

79. "Speckled Trout Fry for Waters of This District," *Porcupine Advance,* March 28, 1929.

80. Frank Evans to Dome Mines Co., June 20, 1922, Dome Mines Co to Frank Evans, June 23, 1922, and Henry Depencier to Alex Fasken, August 14, 1924, file: "Introductions," box 1, F 1250, Dome Mines Company Fonds, Archives of Ontario.

81. "Settler Loses Appeal Against Mining Court," *Porcupine Advance,* 2 October 1930.

82. Joel Tarr, *The Search for the Ultimate Sink: Urban Pollution in Historical Perspective*, 1.

83. Jennifer Gabrys, "Sink: The Dirt of Systems," 679.

84. Lawrence and Davies, "Sludge Question"; Robert Kelley, *Gold vs. Grain: The Hydraulic Mining Controversy in California's Sacramento Valley.*

85. Laurel Sefton MacDowell, "Mining Resources," chap. 5 in *An Environmental History of Canada*, 128–29.

86. See, for example, Robyn Mellor, "A Comparative Case Study of Uranium Mine and Mill Tailings Regulation in Canada and the United States," in *Mining North America*, ed. McNeill and Vrtis, 256–79.

87. Max Liboiron, "Why Discard Studies?"

88. I am drawing here on Max Liboiron, *Pollution Is Colonialism.*

Chapter 5. World of Dust

1. A. R. Riddell, "A Case of Silicosis with Autopsy."

2. Silicosis fits neatly with the conception of slow disaster or slow violence found in environmental history. For example, see Rob Nixon, *Slow Violence and the Environmentalism of the Poor*, 2. "Violence is customarily conceived as an event or action that is immediate in time, explosive and spectacular in space, and as erupting into instant sensational visibility. We need, I believe, to engage a different kind of violence, a violence that is neither spectacular nor instantaneous, but rather incremental and accretive, its calamitous repercussions playing out across a range of temporal scales."

3. See, for example, Christopher Sellers, *Hazards of the Job: From Industrial Disease to Environmental Health Science.*

4. David Rosner and Gerald Markowitz, *Deadly Dust: Silicosis and the Politics of Occupational Disease in Twentieth-Century America*, 14, 15, 18, 19.

5. Rosner and Markowitz, *Deadly Dust*, 21–23.

6. Thomas Oliver, "A Discussion of Miners' Phthisis."

7. Alan Jeeves, *Migrant Labor in South Africa's Mining Economy: The Struggle for the Gold Mines' Labor Supply, 1890–1920*, 46–47.

8. Jock McCullough, "Air Hunger: The 1930 Johannesburg Conference and the Politics of Silicosis," 118.

9. Jock McCullough, *South Africa's Gold Mines and the Politics of Silicosis*, 15.

10. McCullough, *South Africa's Gold Mines*, 18–20.

11. McCullough, *South Africa's Gold Mines*, 38–39.

12. Australia held royal commissions on the dust problem starting in 1905 with the Royal Commission on the Ventilation and Sanitation of Mines in Kalgoorlie. Britain produced the influential 1902–4 Royal Commission into the Health of Cornish Tin Miners, which linked dust and lung disease. Marcus James, "The Struggle Against Silicosis in the Australian Mining Industry: The Role of the Commonwealth Government, 1920–1950," 76–77; McCullough, *South Africa's Gold Mines*, 59.

13. James, "The Struggle Against Silicosis," 77, 78.

14. Rosner and Markowitz, *Deadly Dust*, 25–26, 27.

15. Rosner and Markowitz, *Deadly Dust*, 34–38.

16. Gomez, *Silver Veins, Dusty Lungs.*

17. Todd Gordon and Jeffrey Webber, *Blood of Extraction: Canadian Imperialism in Latin America* and "Canadian Capital and Secondary Imperialism in Latin America"; J. Z. Garrod and Laura Macdonald, "Rethinking 'Canadian Mining Imperialism' in Latin America"; Butler, *Colonial Extractions.*

18. For example, on page 227 of the 1891 (in his inspector's report) Sleight talks about forcing West Silver Mountain Mine to increase ventilation. He does the same for Martindale Mine (Gypsum) on 243. *First Annual Report of the Ontario Department of Mines, 1981* (Toronto: Warwick & Sons, 1892).

19. Price, "Report re. Limitation of the Hours," 11–12.

20. *Twenty-First Annual Report of the Bureau of Mines, 1912,* vol. 21, pt. 1 (Toronto: L. K. Cameron, 1912), 58. The act stated that ventilation was required "so that the shafts, adits, tunnels, winzes, raises, sumps, levels, stopes, cross-cuts, underground stables and working places of the mine and the travelling places to and from such working places shall be in a fit state for working and passing therein." *Annual Report of the Bureau of Mines, 1912,* 64.

21. *Annual Report of the Bureau of Mines, 1912,* 64; "Mining Accidents," *Labor Gazette,* April 1913, 1145.

22. *Twenty-Third Annual Report of the Ontario Bureau of Mines, 1914* (Toronto: L. K. Cameron, 1914), 68–69.

23. *Twenty-Fourth Annual Report of the Ontario Bureau of Mines, 1915* (Toronto: L. K. Cameron, 1915), 78.

24. *Twenty-Second Annual Report of the Bureau of Mines, 1913* (Toronto: L. K. Cameron, 1913), 69.

25. *Annual Report of the Bureau of Mines, 1913,* 70. Corkill's interest in noxious fumes likely stems from the fact that five men died in 1913 from gas remaining in poorly mine shafts following explosives. *Annual Report of the Bureau of Mines, 1913,* 71.

26. *Annual Report of the Ontario Bureau of Mines, 1914,* 69.

27. *Annual Report of the Ontario Bureau of Mines, 1915,* 74.

28. There are no numbers showing whether people actually did this because neither the federal nor the provincial government recorded silicosis in incoming migrants. Furthermore, the blurriness and unpredictability of the speed and moment of "onset" of silicosis made it impossible for physicians to tell when/where a silicotic had contracted it. *Annual Report of the Ontario Bureau of Mines, 1915,* 7474.

29. *Annual Report of the Ontario Bureau of Mines, 1915,* 7478–79.

30. "Issues such as symptomology, what constitutes a disability, what is normal and what constitutes pathology in the human lung, what is an environmental danger, and what is an occupational risk were all questions that every generation in the silicosis debate redefined." Rosner and Markowitz, *Deadly Dust,* 217.

31. Rosner and Markowitz, *Deadly Dust,* 220.

32. "The Risks in Mining," *Globe,* April 16, 1928.

33. G. C. Bateman to H. P. Depencier, May 9, 1933, Silicosis Prior to 1920, box 6, 1925–26 Files, Dome Mines Company Fonds, F 1350, Archives of Ontario.

34. Jabez Elliot, "Silicosis in Ontario Gold Miners," 932, 933, 937.

35. Elliot, "Silicosis in Ontario Gold Miners," 930.

36. Elliot, "Silicosis in Ontario Gold Miners," 931.

37. Elliot, "Silicosis in Ontario Gold Miners," 931.

38. X-ray apparatus receipt, Dome Mines Limited, April 21, 1924, Silicosis, box 10, General Superintendent's Files, 1926–35, Dome Mines Company Fonds, F 1350, Archives of Ontario.

39. In 1929, the equipment was finally moved to Saint Mary's Hospital in Timmins as part of the transfer of silicosis examinations to the mandate of the Worker's Compensation Board that year. "Fine New X-Ray Equipment Installed at the Hospital," *Porcupine Advance,* January 10, 1929.

40. T. F. Sutherland, *Report on the Mining Accidents in Ontario in 1925,* Bulletin No. 54 (Toronto: Clarkson W. James, 1926), 40; "Is to Study Rules Adopted for Mines in Rand of Africa," *Globe,* June 25, 1925.

41. Sutherland, *Mining Accidents, 1925,* 41.

42. Sutherland, *Mining Accidents, 1925,* 41.

43. Sutherland, *Mining Accidents, 1925,* 45–48.

44. T. F. Sutherland, *Report on the Mining Accidents in Ontario in 1926,* Bulletin No. 59 (Toronto: King's Printer, 1927), 12.

45. Sutherland, *Report on Mining Accidents 1926,* 12.

46. Ante-Primary being the least severe, primary and secondary being debilitating, and complicated with tuberculosis being nearly always fatal. Sutherland, *Report on Mining Accidents, 1926,* 11.

47. "Earnest Work in Combatting the Spread of Silicosis," *Porcupine Advance,* July 15, 1926.

48. Minutes of a Meeting of the Silicosis Committee, held at the Hollinger Mine Office, Porcupine, on Wednesday, November 3, 1926, at 8pm, Silicosis, box 10, General Superintendent's Files, 1926–35, Dome Mines Company Fonds, F 1350, Archives of Ontario.

49. "Memorandum in Connection with Examinations by Dr. Haig at Porcupine," Silicosis Prior to 1920, box 6, 1925–26 Files, Dome Mines Company Fonds, F 1350, Archives of Ontario.

50. Meeting of the Silicosis Committee, n.d., Silicosis, box 10, General Superintendent's Files, 1926–35, Dome Mines Company Fonds, F 1350, Archives of Ontario.

51. "Silicosis Grants," *Northern Miner,* March 29, 1928.

52. "Silicosis Not So Bad as Feared," *Northern Miner,* July 14, 1927.

53. "Silicosis Not Bad as Thought," *Northern Miner,* June 21, 1928.

54. John F. Paterson, *Silicosis in Hardrock Miners in Ontario (a Further Study),* Bulletin 173 (Toronto: Ministry of Natural Resources, Ontario, 1973), 33–35.

55. Dr. J. G. Cunningham, "Occurrence of Silicosis in Canada," *Labor Gazette,* January 1931, 39–40.

56. See, for example, C. S. Oettle to the Secretary of Health, July 31, 1929, vol. 629, file 455-13-6, Industrial Health—Diseases—Silicosis, in the Environmental and Occupational Health Files, RG 29, Department of Health Fonds, LAC (hereafter cited as LAC).

57. See, for example, Department of Commerce, Washington, November 18, 1925, vol. 629, file 455-13-6, Industrial Health—Diseases—Silicosis, in the Environmental and Occupational Health Files, RG 29, Department of Health Fonds, LAC; "number and percentages of miners according to nationality and final diagnosis Western Australia—1925–26," vol. 629, file 455-13-6, Industrial Health—Diseases—Silicosis,

in the Environmental and Occupational Health Files, RG 29, Department of Health Fonds, LAC.

58. W. J. Egan to J. A. Amyot, May 14, 1930, vol. 629, file 455-13-6, Industrial Health—Diseases—Silicosis, in the Environmental and Occupational Health Files, RG 29, Department of Health Fonds, LAC.

59. J. D. Page, Chief Immigration Medical Service, to Dr. H. B. Jeffs, October 13, 1930, vol. 629, file 455-13-6, Industrial Health—Diseases—Silicosis, in the Environmental and Occupational Health Files, RG 29, Department of Health Fonds, LAC.

60. Meeting of the Silicosis Committee, n.d., Silicosis, box 10, General Superintendent's Files, 1926–35, Dome Mines Company Fonds, F 1350, Archives of Ontario.

61. Meeting of the Silicosis Committee, n.d.

62. Meeting of the Silicosis Committee, n.d.

63. Dieter Grant Hogaboam, "Compensation and Control: Silicosis in the Ontario Hardrock Mining Industry, 1921–1975," 121.

64. For the American Medical Association recommendation, see "Aluminum in the Prevention and Treatment of Silicosis," *Journal of the American Medical Association* 130, no. 17 (April 1946): 1223. For a discussion, see "Minutes of the Meeting of the Technical Committee on Silicosis," November 3–5, 1948, file 3, container B244379, series F 1352-3, Archives of Ontario; and G. Goralewski, "Clinical and Animal Experimental Studies and the Question of Aluminum Dusty Lung."

65. Memo from W. J. Geldard, May 14, 1958, file 9, container B244380, series F 1352-3, AO.

66. Francis B. Trudeau, "The Objectives and Achievements of the McIntyre Research Foundation," 2.

67. "Minutes of the Meeting of the Technical Committee," April 15–16, 1952, file 5, container 244379, series F1352-3, AO.

68. McIntyre Research Foundation, "Silicosis: What It Is and How It Can Be Prevented," n.d., file 1, container B244379, series F 1352–3, AO.

69. Hogaboam, "Compensation and Control," 121–32.

70. See the April 1, 2019, post on the McIntyre Powder Project Facebook page, https://www.facebook.com/pg/mcintyrepowderproject/posts/?ref=page_internal.

71. Mica Jorgenson and John Sandlos, "Dust Versus Dust: Aluminum Therapy and Silicosis in the Canadian and Global Mining Industries," 4.

72. Dix to James Heller, Executive Administrator, Industrial Disease Standards Panel, November 12, 1990, container 9, series F 4170-7, AO. For a complete history of the McIntyre Foundation through the twentieth century, see Jorgenson and Sandlos, "Dust Versus Dust."

73. "Hollinger System Blamed," *Northern Miner,* May 31, 1928.

74. Rosner and Markowitz, *Deadly Dust,* 8.

Conclusion

1. Maurice Tremblay, "Statistical Review of the Mineral Industry of Ontario for 1938," *Forty-Eighth Annual Report of the Ontario Department of Mines,* vol. 48, pt. 1 (Ottawa: King's Printer, 1940), 11.

2. In total for this period, Canada produced 7,426,000 ounces. Porcupine was responsible for approximately 4,211,500 of those. Figures taken from Tremblay, "Statistical Review," 11; and David Quirin, "Canadian Production of Principal Metallic Minerals, 1886 to 1975," Table P1-26a, Statistics Canada.

3. Different mining regions accomplished their longevity in different ways and according to local geological and political conditions. Daviken Studnicki-Gizbert has written about the longevity of mining in the Sierra Madre, where he argues that, "instead of extending a commodity frontier into untapped regions, capitalist forms of mining in Mexico (and most of Latin America) avoided the pinch of increasing metal scarcity through intensification." Daviken Studnicki-Gizbert, "Exhausting the Sierra Madre: Mining Ecologies in Mexico over the Longue Durée," in *Mining North America,* ed. McNeill and Vrtis, 21.

4. On waste as artifact, see John Baeten, "Contamination as Artifact: Waste and the Presence of Absence at the Trout Lake Concentrator, Coleraine, Minnesota."

5. "Canada Now the Mineral Treasure House to World," *Porcupine Advance,* February 16, 1928.

6. Charles P. Girdwood, Lawrence Jones, and George Lonn, *The Big Dome: Over Seventy Years of Gold Mining in Canada,* 85, 91.

7. Hollinger Annual Report, 1935, 3, Archives of Ontario.

8. Noah Timmins and Sandy McIntyre, among others, died in the early 1930s.

9. W. H. Smith, "Tells the Story of the Discovery of the Hollinger Mines," *Porcupine Advance,* March 7, 1929; "Goldfields to Show New Picture of Hollinger," *Porcupine Advance,* July 11, 1929.

10. "History—Origin of Mine Names—Hollinger, Dome, McIntyre, Coniagas, Norenda etc. Origin of names on T&NO Railway Extension," *Porcupine Advance,* May 8, 1930, 2.

11. "McIntyre Mine—Mine History by R. J. Ennis, General Manager," *Porcupine Advance,* June 15, 1936, "History—Early Mining Facts of Porcupine Revealed by Stephen L'African," *Porcupine Advance,* May 31, 1936.

12. See "Boom Days of Porcupine Returning," *Globe,* August 28, 1924; James Scott, "Who Were Discoverers of Porcupine Camp?," *Globe and Mail,* June 26, 1939.

13. "From Windjammer to Gold Trail: Adventures in Many Countries," *Daily News,* January 15, 1923, http://nla.gov.au/nla.news-article82520086; "Primary Industries of the Empire," *Australasian,* May 28, 1932, 39, http://nla.gov.au/nla.news-article142426668.

14. "Toronto Becomes Big Mining Centre," *Northern Miner,* February 28, 1928.

15. See, for example, B. F. Townsley, *Mine Finders: The History and Romance of Canadian Mineral Discoveries;* and later LeBourdais, *Metals and Men.*

16. A. H. A. Robinson, *Gold in Canada,* 20. By the 1930s, northern Ontario was responsible for more than 90 percent of gold production, and within this context, British Columbia and the Klondike received only passing mention. Robinson, *Gold in Canada,* 19, 23–34.

17. Dominion of Canada Report of the Department of Mines and Resources for the Fiscal Year Ended March 31, 1937 (Ottawa: King's Printer, 1937), 14.

18. Butler, *Colonial Extractions,* 10.

19. For a discussion of the centrality of mining investment for Canada from an Ontario perspective in the postwar era, see Chris Armstrong, *Moose Pastures and*

Mergers: The Ontario Securities Commission and the Regulation of Share Markets in Canada, 1940–1980.

20. Ashifa Kassam, "Guatemalan Women Take on Canada's Mining Giants over 'Horrific Human Rights Abuses,'" *Guardian,* December 13, 2017, https://www.theguardian.com/world/2017/dec/13/guatemala-canada-Indigenous-right-canadian-mining-company; "Barrick Third Quarter 2017 Results," October 25, 2017, http://www.barrick.com/investors/news/news-details/2017/Barrick-Reports-Third-Quarter-2017-Results/default.aspx; "Chile Fines Barrick Gold 16m for Pascua-Lama Mine," BBC News, May 24, 2013, http://www.bbc.com/news/world-latin-america-22663432. For a detailed analysis of Canadian mining and its dubious reputation in Africa, see Butler, *Colonial Extractions.*

21. Gordon Hoeskstra, "Fifth Anniversary Looms with No Charges in Catastrophic Mount Polley Dam Collapse," *Vancouver Sun,* July 7, 2019, https://vancouversun.com/business/local-business/five-year-anniversary-looms-with-no-charges-in-catastrophic-mount-polley-dam-collapse.

22. Nick Ashdown and Niall Mcgee, "In a Turkish Forest, Resistance Grows to a Canadian Company's Gold-Mining Project," *Globe and Mail,* August 26, 2019, https://www.theglobeandmail.com/world/article-in-a-turkish-forest-resistance-grows-to-a-canadian-companys-gold/.

23. Popular historian and Timmins member of parliament Charlie Angus makes a similar argument in *Cobalt.*

24. Butler, *Colonial Extractions,* 10, 11, 61–62.

25. Barbara Hogenboom, "Depoliticized and Repoliticized Minerals in Latin America," 139–40.

26. Bathsheba Demuth makes a similar argument about whales in *Floating Coast: An Environmental History of the Bering Strait,* 46.

27. Most recent figures given by Kam Kotia site-rehabilitation manager Brian McMahon in an interview for CTV News in 2017. "Environmental Cleanup Continues at 'Canada's Worst Mining Disaster' in Timmins," June 8, 2017, CTV News, https://northernontario.ctvnews.ca/environmental-cleanup-continues-at-canada-s-worst-mining-disaster-in-timmins-1.3449983?cache=yzlbeypimu.

Index

About the Author

MICA JORGENSON is an environmental historian specializing in natural resource history, especially gold mining and forestry. She completed her master's in history at the University of Northern British Columbia and her PhD at McMaster University. She has held postdoctoral positions at the Sherman Centre for Digital Scholarship in Canada and as a Marie Skłodowska-Curie Fellow at the University of Stavanger in Norway. She is home when she is in Lheidli T'enneh territory in Prince George, northern British Columbia.